Tobias Meier

Magnetoresistive and Thermoresistive Scanning Probe Microscopy with Applications in Micro- and Nanotechnology

Schriften des Instituts für Mikrostrukturtechnik
am Karlsruher Institut für Technologie (KIT)
Band 25

Hrsg. Institut für Mikrostrukturtechnik

Magnetoresistive and Thermoresistive Scanning Probe Microscopy with Applications in Micro- and Nanotechnology

von
Tobias Meier

Dissertation, Karlsruher Institut für Technologie (KIT)
Fakultät für Maschinenbau

Tag der mündlichen Prüfung: 22. Mai 2014
Hauptreferent: Prof. Dr. Volker Saile
Korreferenten: PD Dr. Hendrik Hölscher, Junior-Prof. Dr. Fabian Pauly

Impressum

Karlsruher Institut für Technologie (KIT)
KIT Scientific Publishing
Straße am Forum 2
D-76131 Karlsruhe

KIT Scientific Publishing is a registered trademark of Karlsruhe
Institute of Technology. Reprint using the book cover is not allowed.

www.ksp.kit.edu

Print on Demand 2014

ISSN 1869-5183
ISBN 978-3-7315-0253-1
DOI   10.5445/KSP/1000042497

Preface

I would like to express my gratitude to the numerous friends, colleagues and institutions for the support and the inspiring discussions during the last years. Without their help, it would have been impossible to finish this thesis. Therefore, it is a pleasure for me to thank them for their contributions and I would like to point out a few for special recognition.

First of all, I want to thank Prof. Dr. Volker Saile for the chance to write this thesis at the *Institute of Microstructure Technology (IMT)* at the *Karlsruhe Institute of Technology (KIT)* and his scientific and personal support for my work during this time. Furthermore, I like to acknowledge the great support of Prof. Dr. Juerg Leuthold and Prof. Dr. Uli Lemmer.

Specifically, I wish to highlight the excellent support during this work by my supervisor PD Dr. Hendrik Hölscher. Due to both, his interpersonal and professional regard, it was a pleasure to discuss and tackle the exciting challenges on which this work is based.

Furthermore, I'd like to thank the colleagues at Christian Albrechts University Kiel, Ali Tavassolizadeh, Dr. Dirk Meyners and Prof. Dr. Eckhard Quandt, and at Bielefeld University, Dr. Karsten Rott and Prof. Dr. Günter Reiss, who fabricated the magnetoresistive tunneling structures and incorporated them into AFM cantilevers. Without their continuously hard work, on fabricating AFM cantilevers, the results presented on magnetoresistive self-sensing cantilevers would not have been possible. At this point, I acknowledge the financial support for this project from the *Deutsche Forschungsgemeinschaft (DFG)*. Additionally, I want to thank Dr. Roland Gröger, Dr. Alexander Förste and Dr. Stefan Walheim from the *Institute of Nanotechnology (INT)* at the *Karlsruhe Institute of Technology (KIT)* for the fruitful discussions during the design and set up of the specialized instrument for the implementation of the magnetoresistive AFM cantilevers.

I am very much indebted to Dr. Bernd Gotsmann of *IBM Research - Zurich* for inviting me to his laboratory and introducing me to the concept of scanning thermal microscopy. It was a great pleasure to work with him and exciting to be part of this highly innovative group. I further want to express special gratitude to Fabian Menges, who did an amazing job on setting up the scanning thermal microscope at the *IBM labs*. For funding my research period at *IBM Research - Zurich*, I would like to thank the *Karlsruhe House of Young Scientists (KHYS)*.

It is also a pleasure to thank my colleagues of the *Nano- and Micoreplication Group* at the *Institute of Microstructure Technology (IMT)* for the good cooperation and working atmosphere during the processing of shape memory polymers. I wish to express my gratitude to PD Dr. Matthias Worgull, Marc Schneider and Dr. Alexander Kolew for advice, technical help and most of all inspiring discussions and a helping hand whenever needed. In this regard, I would like to acknowledge the support of the *Karlsruhe Nano Micro Facility (KNMF)*, a Helmholtz Research Infrastructure at *Karlsruhe Institute of Technology (KIT)*.

Additionally, I want to thank Paul Abaffy for the numerous hours at the scanning electron microscope. In addition, I wish to thank the entire workshop team at IMT and Andreas Deck from the workshop at the *Institute of Applied Physics*. At this point, I also want to thank my office mates Taleieh Rajabi, Christian Lay, Radwanul Siddique and Michael Röhrig - it was a pleasure working with you.

It is also a pleasure to acknowledge fruitful discussions with Dr. Sven Schüle, Dr. Markus Simon, Dr. Zhenhao Zhang, Dr. Maryna Kavalenka, Dr. Julia Syurik, Senta Schauer, Richard Thelen, Kira Köhnle, Dr. Klaus Feit, Markus Heilig, Norbert Schneider, Oliver Krömer and Benjamin Leyrer at KIT. Outside KIT, I would like to thank Dr. Heike Riel, Dr. Peter Nirmalraj and Dr. Heiko Wolf at IBM Research - Zurich; Klaus Pross, Johannes

Kindt and Steve Minne of Bruker Nano Surfaces Division; Friedhelm Freiss, Stephan Vinzelberg, Amir Moshar and Ted DuPar of Asylum Research for inspiring discussions and their friendly welcome at their laboratories.

Finally, I would like to thank my family, especially my parents and Nicole, for their understanding and loving support during the last years.

Karlsruhe, April 2014

Abstract

In modern micro- and nanotechnology, knowledge from the macroscopic world has to be questioned whether it still applies at the micro- and nanoscale. Aiming for smaller structural feature sizes and smart usage of these new material properties can improve current devices and pioneer the technological platform for new applications. The key to this technology can be found in the ability to image samples on the nanoscale. This work presents approaches to extend limits of scanning probe microscopy techniques towards more versatile instruments. Integrated sensor concepts by magnetoresistive and thermoresistive sensing are presented. Furthermore, a fabrication method to design application-inspired micro- and nanostructures is introduced.

To combine the advantages of high resolution scanning probe microscopy with a large field of view, a new atomic force microscope with a nested scanner design was developed. This unique microscope benefits from two independent scanners, one with a large range of $800 \times 800 \, \mu m^2$ and one with a scan range of $5 \times 5 \, \mu m^2$ for high spatial resolution.

Additionally, the instrument is designed to be operated with both, a conventional beam deflection setup and self-sensing cantilevers based on magnetic tunneling junctions with magnetostrictive electrodes. This concept showed sufficient sensitivity to be used in atomic force microscopy and even outperformed specifically optimized piezoresistive and piezoelectric cantilevers.

Utilizing thermoresistive cantilevers in scanning thermal microscopy, the heat flux between the tip of the cantilever and the sample was measured with high precision. This sensor concept enabled quantitative thermal conductance measurements of linear molecular chains as a function of their chain length. Self-assembled monolayers of various linear alkanes were studied as a model system and showed signatures of phonon localization and interference.

The switching properties of thermally triggered shape memory polymers were analyzed in micro- and nanoscale systems. These materials were introduced to the fabrication of molds for the replication of micro- and nanoscale components. Utilizing the shape memory effect for replication of micro- and nanostructures, allowed the fabrication of self-healing and demolding molds for the replication of micro- and nanostructures on curved surfaces.

Kurzfassung

In der modernen Mikro- und Nanotechnologie hängen Materialeigenschaften nicht nur von deren Zusammensetzung ab, sondern auch stark von deren Größe und Form. Das Ziel mit immer kleineren Strukturelementen neue Materialeigenschaften intelligent zu nutzen, führt dabei nicht nur zur Verbesserung gängiger Produkte, sondern kann den Weg zu neuen Anwendungen und Technologieplattformen bereiten. Die Fähigkeit, Proben auf dieser Größenskala abzubilden, öffnet dabei die Tür zu dieser Technologie. Diese Arbeit zeigt daher Wege auf, um die bereits große Familie der Rastersondenmikroskope noch vielseitiger zu gestalten. Dabei wurde zunächst die Abtastfläche von typischerweise $100 \times 100\,\mu m^2$ auf $800 \times 800\,\mu m^2$ mit einem Dual-Scanner System vergrößert. Das Dual-Scanner System verfügt zudem über einen unabhängigen $5 \times 5\,\mu m^2$ Scanner für eine hohe lokale Auflösung.

Zusätzlich wurde das Mikroskop für die gleichzeitige Nutzung eines konventionellen Lichtzeiger Detektors und dem Einsatz neuartiger selbstdetektierender Mikrofederbalken mit integrierten magnetischen Tunnelsensoren ausgelegt. Diese Tunnelsensoren mit magnetostriktiven Elektroden zeigten eine herausragende Sensitivität für den Einsatz in der Rasterkraftmikroskopie und erwiesen sich piezoelektrischen und piezoresistiven Sensoren, welche speziell für diese Anwendung optimiert wurden, überlegen.

Mittels thermoresistiver Mikrofederbalken wurden hochpräzise thermische Leitwertmessungen an nanoskalaren Probensystemen durchgeführt. Durch dieses Sensorkonzept konnten quantitative Messungen des thermischen Leitwertes von linearen Alkanketten als Funktion ihrer Kettenlänge innerhalb eines Modellsystems von selbst organisierter Monolagen durchgeführt werden. Dieses Modellsystem zeigte starke Anzeichen von Frequenzeingrenzung und Interferenz von Phononen entlang der Alkanketten.

Zuletzt wurden die Schalteigenschaften thermisch aktivierbarer Formgedächtnispolymere untersucht. Durch deren Einsatz in der Herstellung von Formeinsätzen für die Mikro- und Nanoreplikation, wurde eine neue Klasse von intelligenten Formwerkzeugen vorgestellt. Mit der Einführung dieser neuartigen Werkzeuge konnten erstmals Mikro- und Nanostrukturen auf gekrümmten Vollmaterialoberflächen sowie mit Hinterschnitten entformt werden.

Contents

1. Introduction

When length scales known from daily life are left behind, the applicability of knowledge from the macroscopic world must also be questioned. In modern micro- and nanotechnology, the knowledge from the macroscopic world cannot be transferred to the micro- and nanoscale. New material properties can improve current devices and pioneer the technological platform for new applications as the feature size becomes a relevant factor in nanosystems. In classical solid state physics, describing the bulk only is a good approximation of the macroscopic material properties: surface effects can be neglected, greatly simplifying the analytic description. On the nanoscale, this is not the case. To illustrate the influence of the surface in micro- and nanoscale systems, the surface to volume ratio of two cubes, one with an edge length of 1 µm and one cube width an edge length of 10 nm filled with densely packed atoms of 3 Å diameter, can be calculated. For the larger cube, only 0.2 % of the atoms are on its surface while 2 % of the atoms of the smaller cube are surface atoms. By minimizing the feature sizes, surface effects even start to dominate the material properties while well known effects of the bulk material are suppressed.

As Richard Feynman pointed out in his visionary talk "There's Plenty of Room at the Bottom"[1] the key to the nanoscale world can be found in microscopy techniques. As conventional optical microscopes are limited by the diffraction limit,[2] not only the expected new material properties on the nanoscale are a challenge to explain, also observing them was a challenge in the 1960s and, in some cases, is still today. Introducing a new class of microscopes, the invention of the scanning tunneling microscope (STM)[3] by Binnig *et al.* in 1981 and of the atomic force microscope (AFM)[4] in 1986 has provided not only ground breaking microscopy techniques but also versatile tools for manipulation of single atoms.[5] Nowadays, the atomic force microscope is one of the most widespread tools and one of the workhorses in nanotechnology laboratories. However, multiple sensor concepts to detect interactions between a tip and sample are competing and adapted to specific applications.

In the first part of this work, limitations of current instruments are discussed. A new scanning probe microscope using a nested scanner design enabling both, a unique large scan area and a high spatial resolution is designed. By combining large area scanning probe microscopy and optical microscopy in a single instrument, a versatile tool for micro- and nanoscale surface analysis is presented

Even as the atomic force microscope is a well established tool for modern nanotechnology, still most instruments rely on optical read-outs of micro-machined silicon cantilevers. Optical read-outs, however, require bulky mechanical alignment components and their implementation is often a challenge in specific environments like vacuum or liquids. Additionally, the optical read-out can also influence the cantilevers deflection[6] or interfere with the sample.[7] To overcome these issues, cantilevers with integrated sensing elements, so called self-sensing cantilevers, based on magnetic tunneling junctions with magnetostrictive electrodes are introduced.[8] The strain sensitive magnetic tunneling junctions can be integrated onto micro-machined cantilevers and allow static as well as dynamic operational modes of AFMs with sufficient sensitivity to resolve atomic step-edges.

By implementing of thermoresistive sensors[9,10] into micro-machined silicon cantilevers, the atomic force microscope is converted into a scanning thermal microscope,[11] which is highly sensitive to heat flux between the tip and the sample.[12,13] This can be utilized to investigate heat transfer on the nanoscale. In the third section of this work, thermal transport measurements along one dimensional alkane chains as a function of their length is presented. Even as self-assembled monolayers of linear alkane chains are of high technological relevance, their thermal properties still remain unclear even as one dimensional systems have been discussed for more than a century.[14, 15] Applying a nanoscopic microscopy technique to such systems provides experimental insight for the ongoing discussion.

In the last part of this work, material properties of thermally triggered shape memory polymers are investigated for the application in micro- and nanoscale systems. The investigated thermoplastic smart materials are capable of defined shape changes down to the sub-micrometer range. Utilizing those polymers for the fabrication of molds for the replication of micro- and nanostructures allows the fabrication of components with a smart and three dimensional surface morphology. By using shape memory polymeric molds, an additional degree of freedom to the design of polymeric micro- and nanoscale devices is introduced.

2. Experimental Methods

In the past decades, various fabrication and microscopy techniques have been developed for modern micro- and nanotechnology. After first conceptional origins[1] and first concepts of molecular assembly of nanosystems,[16, 17] the invention of the scanning tunneling microscope (STM)[3] by Binnig *et al.* in 1981 and of the atomic force microscope (AFM)[4] in 1986 provided a unique way to observe and manipulate even atom directly.[5] Nowadays, the atomic force microscope is the most widespread tool for nanotechnology and is not only used in research laboratories, but also for routine quality control and product development in many companies. This chapter introduces the basic concepts of scanning probe microscopy, nanomechanics, fabrication methods and materials.

2.1. Principles of Scanning Probe Microscopy

Like Feynman pointed out in his famous talk, the keys to micro and nanotechnology are microscopy techniques.[1] In 1881, Ernst Abbe found that classical wave optical techniques are limited by the diffraction limit, which is around 200 nm for visible light.[2] Microscopy techniques, which can achieve a higher resolution are scanning probe based methods. They are based on the simple and Nobel Prize winning idea of Binnig and Rohrer[3] to scan an atomically sharp tip across a sample and reconstruct the samples topography by the interactions between the tip and the sample. The first scanning probe microscope, the scanning tunneling microscope was measuring the electric tunneling current between tip and sample and was, therefore, limited to conductive samples. The further development of the STM to the atomic force microscope overcame those limitations by using a mechanical force sensor to sens tip-sample interactions. By measuring the force interactions between tip and sample, the AFM can achieve a sensitivity sufficient to reveal intramolecular structures[18] and the arrangement of single atoms.[19]

Since interatomic force are very small and in the range of some nN, the force sensor of an AFM has to be very precise.[20] In order to sens forces in an AFM with this high precision, the tip is mounted at the end of a bendable cantilever which acts as a spring. The typical stiffness of the spring is, depending on the operational mode of the AFM, between $0.01\,\mathrm{N/m}$ and $100\,\mathrm{N/m}$. The deflection of the cantilever Δz with a spring constant k_z is directly proportional to the applied force F_z. As the cantilever is deflected in the elastic regime, the force can be described by Hooke's law:

$$F_z = k_z \cdot \Delta z \tag{2.1}$$

To detect small forces, the deflection of the cantilever must be measured very precisely. However, since the tip-sample forces are in the nN-range, the cantilever is deflected between $0.01\,\mathrm{nm}$ and $100\,\mathrm{nm}$. This requires a very precise measurement of the cantilever deflection on these small scales. As shown in Fig. 2.1, multiple detection methods and sensing elements for scanning probe cantilevers are available and can be optimized for specific applications. In this work, the well established laser beam deflection method[21] was used and compared to new detection methods like magnetoresistive sensing[8] and thermoresistive sensing.[13] A detailed description of the used sensing mechanisms can be found in Chap. 3.

To understand the interaction between tip and sample and the mechanics of nanoscale contacts, the large variety of tip-sample forces and contact mechanics have to be described and are summarized in the following.

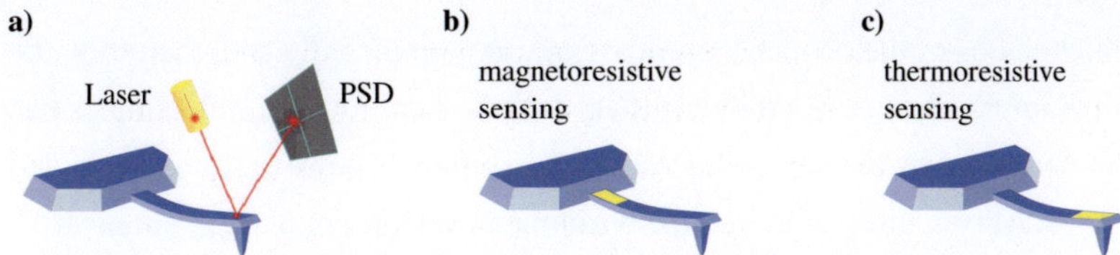

Figure 2.1.: The key elements in atomic force microscopes and scanning thermal microscopes are sharp tips on microfabricated cantilevers. The forces between tip and sample are measured by the deflection of the cantilever. **a)** A laser beam deflection set-up is the most widely spread method to detect the cantilever deflection. By focusing a laser beam on the backside of the cantilever, the deflection can be measured by monitoring the reflected beam with a position sensitive photo detector. **b)** An other method is to measure the strain in the cantilever caused by its deflection. A magnetoresistive tunneling sensor can offer a high strain sensitivity and can be integrated into the cantilever during the fabrication process at the cantilever base. Magnetoresitive cantilevers can therefore greatly simplify the set-up. **c)** Scanning thermal microscopy focuses on measuring the heat flux from the tip into the sample instead of tip-sample forces. Thermoresistive sensors offer a high sensitivity to heat flux and are integrated on the base of the tip.

2.1.1. Tip-Sample-Forces

By approaching a tip to a surface, various forces start to act on the tip apex at small distances. Those forces can be attractive or repulsive and are strongly distance dependent. Depending on the actual tip-sample distance, different forces become dominant. The geometry of tip and sample in atomic force microscopy is typical approximated by a spherical tip and a flat sample.

Capillary Forces

In ambient conditions, water condenses on all surfaces. The thickness of these water layers, which can reach between 10 nm and 200 nm, depends on the humidity and temperature.[22] Around and within the tip-sample contact, a water meniscus with very strong adhesive forces is formed.[23] Water films can, therefore, dramatically influence the image quality. On samples with superhydrophilic and superhydrophobic surface properties, the film thicknesses of the condensed water can vary significantly and, therefore, the tip-sample forces caused by capillary forces. Additionally, soft samples can be deformed by the strong forces. Capillary forces, however, can be avoided by measuring directly in liquid environment, vacuum or dry gas atmosphere.

Van der Waals Forces

On shorter length scales, van der Waals forces become dominant. The origin of van der Waals forces are induced electric dipoles in neutral atoms and molecules. Even if the atom or molecule has no initial dipole moment, a dipole moment can be induced on short timescales as the distribution of charges is statistically and must not be symmetric on those short timescales. To simplify the geometry of tip and surface, a sphere with radius R approaching a flat surface is a good approximation. For this geometry, the van der Waals force as a function of distance z is given by[24]

$$F_{vdW}(z) = -\frac{A_H R}{6z^2} \tag{2.2}$$

where A_H is the Hamaker constant[25] with values around $10^{-19}\,\mathrm{J}$.[24] Van der Waals forces are attractive forces as indicated by the negative sign. As they are proportional to $1/z^2$, they are considered as long range forces compared to other forces occurring in AFM experiments.

Pauli Repulsion

Postulated by Wolfgang Pauli in 1925, the Pauli exclusion principle forbids significant overlap of two charge clouds with two electrons in the same quantum mechanical state.[26,27] This work inspired Goerge Uhlenbeck and Samuel Goudsmit to introduce the spin of electrons[28] increasing the principle quantum numbers from three to four.[29] The Pauli repulsion describes a strong repulsive force, which acts when the tip is in contact with the surface. Additionally, the Pauli exclusion principle also effects the charge distribution of the electrons in the tip's and sample's atoms in a way, that the shielding of the nucleus charges is reduced. This causes an additional ionic repulsion. As a result, the Pauli and ionic repulsion are very strong and dominating while the tip is in contact with the sample. As the Pauli repulsion is purely quantum mechanical and the ionic repulsion can be described with Coulombs law, the description with empirical potentials like the Lennard-Jones potential provides an easy and fast calculation of repulsive and attractive forces[24]

$$V_{\mathrm{LJ}}(z) = E_0\left(\left(\frac{r_0}{z}\right)^{12} - 2\left(\frac{r_0}{z}\right)^{6}\right)$$

(2.3)

The Lennard-Jones potential takes also attractive van der Waals forces into account. It can be defined with the bonding energy E_0 and an equilibrium distance r_0. For the 12-6 potential given in Eq. (2.3), the repulsive forces are represented by the term with the inverse power law with $n = 12$ and the attractive forces by the term with $n = 6$. This potential can be used to describe the forces between individual atoms/molecules. However, an atomic force microscopy tip and also all the samples are not single atoms and molecules but ensembles of atoms. Therefore, also other effects like elastic deformation of tip and sample have to be taken into account.

Contact Mechanics

If the tip is in contact with the sample, the tip applies a load to the sample. The applied load causes an elastic deformation of tip and sample. This deformation effects the effective contact area and is an important issue in atomic force microscopy. A schematic of the contact geometry for a layered sample can be found in Fig. 2.2 a). The elastic deformation of two spheres and resulting repulsive forces were first analyzed by Heinrich Hertz in 1881.[30,31,32] The Hertz model neglects adhesion forces and can, therefore, only be used for low adhesive samples. For atomic force microscopy experiments, the original Hertz model has to be modified to a flat-spherical geometry. The repulsive force for the contact regime in this flat-spherical Hertz model is

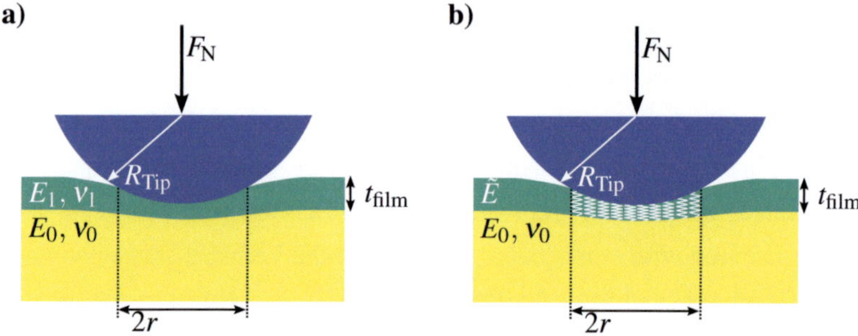

Figure 2.2.: Indentation model of a layered sample with a rigid sphere. **a)** In the actual configuration, the film (with the Young's modulus E_1 and Poisson's ratio v_1) is assumed to be perfectly bound to the substrate (with the Young's modulus E_0 and Poisson's ratio v_0) while the contact interface is frictionless. **b)** With these assumptions, the layer can be modeled as a mattress transmitting normal stress into the substrate with its effective compression modulus $\tilde{E}$ being its oedometric modulus.

$$F_{\text{Hertz}}(z) = \frac{4}{3}\tilde{E}\sqrt{R}(z_0 - z)^{3/2} \qquad \text{for } z \leq z_0 \tag{2.4}$$

with the effective elastic modulus $\tilde{E}$

$$\tilde{E} = \frac{E_t E_s}{E_s\left(1 - v_t^2\right) + E_t\left(1 - v_s^2\right)}. \tag{2.5}$$

The Hertz force depends on the tip radius R as well as the point of contact z_0, the Young's moduli $E_{t,s}$ and the Poisson rations $v_{t,s}$ of tip and sample, respectively. This results in a contact radius of

$$r_{\text{Hertz}} = \left(\frac{R F_{\text{Hertz}}}{\tilde{E}}\right)^{1/3}. \tag{2.6}$$

Extending the theory of Hertz to adhesive contacts has been done by Johnson, Kendall and Roberts (JKR)[33] as well as by Derjaguin, Muller and Toporov (DMT).[34] The JKR theory only takes adhesion within the contact area into account while the DMT theory considers also long range adhesive forces outside the contact.

The JKR theory models the adhesion force by a delta function γ representing the surface energy at the contact distance z_0. During approach of the tip towards the sample, this implies that all interactions for distances larger than z_0 vanish and the adhesive force has a infinitely short range. For the surface energy, this requires a change of $U_S = -\pi r^2 \gamma$. As in the Hertzian contact model, where the applied load and tip-sample force are identical, the JKR-tip-sample force is offset by the adhesion force which pulls the surface into contact over an area which exceeds that given by the Hertz theory.[32] This results in a effective Hertzian tip-sample force of[35]

$$F_{\mathrm{JKR}}(z) = F_{\mathrm{N}}(z) + 3\pi R \gamma \pm \sqrt{6\pi R \gamma F_{\mathrm{N}}(z) + (3\pi R \gamma)^2} \qquad (2.7)$$

$$= \left(\sqrt{-F_c^{\mathrm{JKR}}} \pm \sqrt{F_{\mathrm{N}}(z) - F_c^{\mathrm{JKR}}} \right)^2. \qquad (2.8)$$

$F_c^{\mathrm{JKR}} = -1.5\pi R \gamma$ represents the so-called critical force, the force needed to separate tip and sample while the negative solution of Eq. (2.8) denotes unstable conditions. The contact radius in this model is

$$r_{\mathrm{JKR}} = \left(\frac{R}{\tilde{E}} \right)^{1/3} \left(\sqrt{-F_c^{\mathrm{JKR}}} \pm \sqrt{F_{\mathrm{N}}(z) - F_c^{\mathrm{JKR}}} \right)^{2/3} \qquad (2.9)$$

As this model neglects adhesion forces outside the contact area, it describes the contact mechanics well for situations where the forces within the contact area are dominant. This is the case, when the contact area becomes very large as it is the case for large tip radii or soft sample materials like soft polymers or biological tissues.

If the assumptions of the JKR model do not apply, like it is the case for small tip radii and rigid samples, the DMT theory[34,35] describes the contact mechanics on the other extreme case very well. The analysis of Derjaguin, Muller and Toporov include adhesion forces outside the contact area like long range van der Waals forces, however their original theory did not result in an analytic solution. Maugis proposed in 1992 an approximation to the original DMT theory, which is often referred as the DMT-M or *Hertz-plus-offset model*.[36] In this model, the elastic forces between tip and sample are calculated like in the Hertz theory, but with a offset taking the adhesion forces into account. Maugis proposed following effective Hertzian load

$$F_{\text{DMT-M}}(z) = F_{\text{Hertz}}(z) - F_c^{\text{DMT}} \tag{2.10}$$

$$= \frac{4}{3}\tilde{E}\sqrt{R}(z_0 - z)^{3/2} - F_c^{\text{DMT}}. \tag{2.11}$$

$F_c^{\text{DMT}} = -2\pi R\gamma$ is in this theory the critical force of the DMT model with $\gamma = \frac{A_H}{12\pi z^2}$. The contact radius, consequently, can be expressed like in the Hertz model with the modified force

$$F_{\text{DMT-M}}(z) = \begin{cases} -\dfrac{A_H R}{6z^2} & \text{for} \quad z \geq z_0\,, \\ \frac{4}{3}\tilde{E}\sqrt{R}(z_0 - z)^{3/2} - \dfrac{A_H R}{6z_0^2} & \text{for} \quad z < z_0\,. \end{cases} \tag{2.12}$$

$$r_{\text{DMT-M}} = \left(\frac{R F_{\text{DMT-M}}}{\tilde{E}}\right)^{1/3} \tag{2.13}$$

In Fig. 2.3 the tip-sample forces resulting of the proposed DMT-M contact model is given.

Thin Film Compression Model

For the contact mechanics not only the understanding of tip-sample forces is a precondition, but also also a description of the sample is necessary. As discussed above, the Young's modulus of the sample is an important parameter to describe the contact mechanics. For a layered sample system (substrate + thin film) like the one used in the experiments conducted in Chap. 5, the effective modulus can vary with the modulus of the film material. For thin films, a model introduced by Johnson[32] can be used to calculate the contact area. It has shown to be accurate for a tip-sample system with

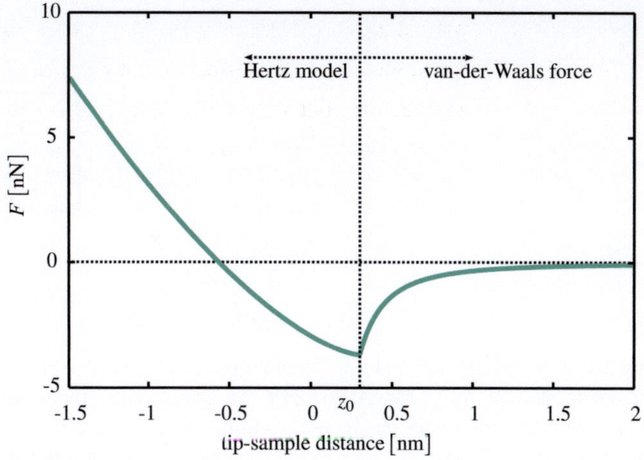

Figure 2.3.: Realistic tip-sample force by the DMT-M model. The parameters used for the plot represents typical values for atomic force microscopy experiments under ambient conditions: $A_H = 0.2\,\text{aJ}$, $R = 10\,\text{nm}$, $E_t = 130\,\text{GPa}$, $E_s = 1\,\text{GPa}$ and $v_t = v_s = 0.3$, $z_0 = 0.3\,\text{nm}$.

sufficiently large contact radii.[37,38,39] According to this model, on samples with a relatively hard substrate covered with a soft film, ($E_0 \gg E_1$ with the Young's modulus E_0 of the substrate and the Young's modulus E_1 of the film, respectively) an effective modulus $\tilde{E}$ can be defined (see Fig. 2.2). The effective modulus of the film can be expressed by the oedometric approximation using the Poisson's ratio v_1 of the film.

$$\tilde{E} = E_1 \frac{1 - v_1}{(1 - 2v_1)(1 + v_1)} \tag{2.14}$$

When shear forces can be neglected and the DMT-M model is applied to consider adhesion, the contact radius r is given by[37]

$$r = \left(\frac{4tR_{\text{Tip}}}{\pi\tilde{E}} F_N \right)^{1/4} \tag{2.15}$$

This analytical model gives a good approximation of the contact radius if the contact radii r is large compared to the thickness t of the film, $r/t > 1$.

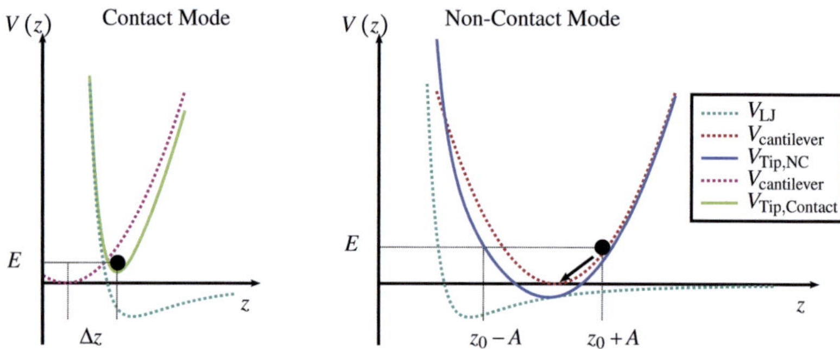

Figure 2.4.: Tip-sample potentials for contact mode and non-contact mode. In contact mode, the tip is moved towards the sample until the repulsive force are dominant for better mechanical stability of the measurement. In amplitude modulation mode, the tip oscillates with its resonance frequency in the harmonic oscillator potential. As the tip approaches the sample, first long range forces distort this potential changing resonance frequency, amplitude and phase of the oscillation.

The normal force can be calculated with the tip-sample forces expressed by the models described above.

2.1.2. Operation Modes of Atomic Force Microscopes

Atomic force microscopes can be operated in various operational modes. The two most important ones are contact mode and non contact mode. In order to describe contact and non contact mode, the cantilever's deflection can be modeled by an interplay of a tip-sample potential given by Eq. (2.3) and the harmonic oscillator potential $V_{\text{spring}}(z) = \frac{k}{2}(z - z_0)^2$ of the cantilever.[40]

The contact mode is historically the oldest operational mode. In this mode, this tip is brought into contact with the sample. As the tip approaches to the sample, the tip is statically trapped in the minimum of the effective tip-sample potential and the cantilever is deflected statically. However, as shown in Fig. 2.4, depending on the tip-sample distance, the potential minimum is on different locations corresponding to the z-position of the cantilever. This raises the question of the mechanical stability of the measurement. As long as the long range attractive forces are dominant, the restoring force of the cantilever counteracts the tip-sample force in a way that it tends to separate tip and sample. If the force gradient of the tip-sample force is larger than the spring constant, the tip will "snap" towards the surface. Mathematically expressed, this instability occurs if[41]

$$k_z < \frac{\partial F_{\text{ts}}(z)}{\partial z} \tag{2.16}$$

For mechanical stable measurement, avoiding the unstable regime is an advantage. The z-position of the tip is, therefore, adjusted to a position where the force gradient is either small or where the tip-sample force is counteracting the force applied by the cantilevers spring constant.

In contact mode, the repulsive forces are the dominant tip-sample forces. The tip is pushed onto the sample and the cantilever is deflected by Δz and the force $F = k_z \times \Delta z$ acts on the sample. The effective potential of the tip-sample forces and the cantilever shows a single minimum (see Fig. 2.4). As this minimum is above the minimum of the harmonic potential, the tips position is stable at the minimum of the effective potential which is the equilibrium position of the interplay of tip-sample force and the applied force by the cantilever. In other words, the tip is pressed onto the sample for mechanical stability. Due to this physical contact, however, friction forces occur during scanning of the tip and can be measured by the torsion of the cantilever.[42] But friction also leads to wear and can change the tip geometry or damage the sample.

To minimize wear and sample damage, controlling the force between tip and sample is done with a feedback loop (see Fig. 2.5). The desired applied force by the cantilever is given to the feedback loop as a set-point. The feedback loop controls the sample z-position for a constant tip-sample force.

Even as contact mode AFM can perform measurements with high mechanical stability, friction forces often limit the resolution (especially on soft samples). To avoid friction forces, dynamic operational modes can

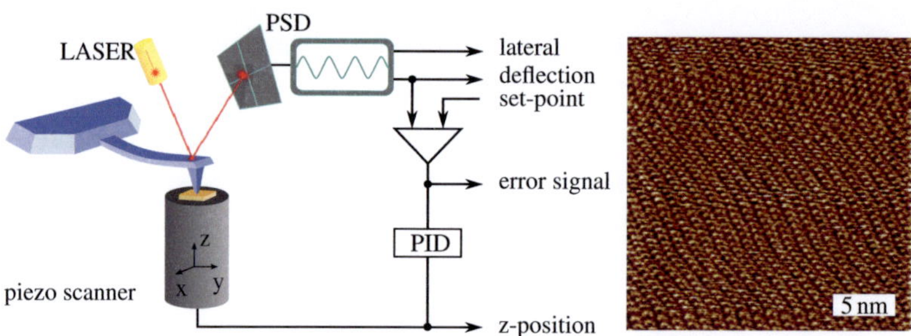

Figure 2.5.: Principle of the feedback loop of the atomic force microscope in contact mode. The cantilever deflection is measured with a laser beam deflection set-up. To control the tip-sample force, the desired applied force is given as set-point to a feedback loop adjusting the sample position in a way that the deflection is kept constant. If the AFM is operated in dynamic modes like amplitude modulation, atomic resolution can be reached with such setups. On the right, an example of atomically resolved calcite is given.

be used to scan the sample while the tip is oscillated in a distance close to the sample, where the force gradient is small. The basic idea of this operational mode is to keep the tip in the non contact regime, where the long range forces are dominant.[43] As this regime show some instabilities, the measurement becomes more stable if not the static deflection is measured like in contact mode, but the tip is oscillating in the effective potential. The distorted harmonic oscillator potential of the cantilever determines thereby the motion of the tip $z(t)$. The free resonance frequency f_0 of the oscillation is determined very far away from the surface. This ensures that only the harmonic potential is dominating the resonance frequency. While driven next to the cantilevers resonance frequency, the tip is approached towards the sample. Close to the sample, the tip-sample potential distorts the harmonic potential of the cantilever and the resonance frequency of the oscillation changes. This has also effects on the amplitude and phase response of the oscillation. The movement of the tip in this distorted harmonic oscillator potential can be described by the equation of motion

$$m\ddot{z}(t) + \frac{2\pi f_0 m}{Q_0}\dot{z}(t) + k_z(z_0 - z(t)) = \underbrace{A_d k_z \cos(2\pi f_d t)}_{\text{external driving force}} + \underbrace{F_{ts}\left[z(t),\dot{z}(t)\right]}_{\text{tip-sample force}}$$

$$(2.17)$$

In this notation, $z(t)$ is the tip position at the time t. The parameters k_z, m, z_0 and $f_0 = \sqrt{k_z/m}/(2\pi)$ are the spring constant, effective mass, equilibrium position and the eigenfrequency of the cantilever. The quality factor of the resonance Q_0 includes intrinsic damping of the cantilever and of the surrounding media for simplification. The cantilever is driven with an amplitude A_d and frequency f_d. The most widespread dynamic mode is amplitude modulation, where A_d and f_d are kept constant.

Driven far away from the sample, tip-sample forces F_{ts} can be neglected and the cantilever oscillation follows the well-known equation of motion of a driven-damped harmonic oscillator. A steady-state solution can be given by the ansatz

$$z(t = 0) = z_0 + A_0 \cos(2\pi f_d t + \varphi) \qquad (2.18)$$

with the phase difference φ between the excitation and the cantilever oscillation. Using the ansatz from Eq. (2.18) to solve the equation of motion (Eq. (2.17)), the amplitude A_0 and phase φ can be expressed by

$$A_0 = \frac{A_d}{\sqrt{\left(1 - \frac{f_d^2}{f_0^2}\right)^2 + \left(\frac{1}{Q_0}\frac{f_d}{f_0}\right)^2}} \qquad (2.19)$$

$$\tan\varphi = \frac{1}{Q_0}\frac{\frac{f_d}{f_0}}{1 - \frac{f_d^2}{f_0^2}} \qquad (2.20)$$

If the cantilever approaches the surface, the tip-sample force $F_{ts}\left[z(t),\dot{z}(t)\right]$ has to be considered. The highly non-linear tip-sample potential complicates the analytic solution of Eq. (2.17). Hölscher *et al.*[44] proposed to focus on the steady state solutions of the equation of motion with sinusoidal cantilever oscillation and expand the tip-sample force into its Fourier series

$$
\begin{aligned}
F_{\text{ts}}\left[z(t),\dot{z}(t)\right] \approx\ & f_{\text{d}} \int_0^{1/f_{\text{d}}} F_{\text{ts}}\left[z(t),\dot{z}(t)\right] dt \\[6pt]
& + 2f_{\text{d}} \int_0^{1/f_{\text{d}}} F_{\text{ts}}\left[z(t),\dot{z}(t)\right] \cos\left(2\pi f_{\text{d}}t + \varphi\right) dt \times \cos\left(2\pi f_{\text{d}}t + \varphi\right) \\[6pt]
& + 2f_{\text{d}} \int_0^{1/f_{\text{d}}} F_{\text{ts}}\left[z(t),\dot{z}(t)\right] \sin\left(2\pi f_{\text{d}}t + \varphi\right) dt \times \sin\left(2\pi f_{\text{d}}t + \varphi\right) \\[6pt]
& + \ldots
\end{aligned}
\tag{2.21}
$$

Inserting the first harmonics of Eq. (2.21) into the equation of motion Eq. (2.17), Hölscher *et al.*[44] found two coupled equations

$$
\frac{f_0^2 - f_{\text{d}}^2}{f_0^2} = I_+\left(z_0,A\right) + \frac{A_{\text{d}}}{A}\cos\varphi
\tag{2.22}
$$

$$
-\frac{1}{Q_0}\frac{f_{\text{d}}}{f_0} = I_-\left(z_0,A\right) + \frac{A_{\text{d}}}{A}\sin\varphi
\tag{2.23}
$$

Here, the integral I_+ is a weighted average of the tip-sample forces during approach $F_{\Downarrow}$ and retract $F_{\Uparrow}$ of the tip within one oscillation. The integral I_- is directly connected to the dissipated energy ΔE during one oscillation cycle. The integrals I_+ and I_- are[44]

$$I_+(z_0,A) = \frac{2f_\mathrm{d}}{k_z A} \int_0^{1/f_\mathrm{d}} F_\mathrm{ts}\left[z(t),\dot{z}(t)\right] \cos\left(2\pi f_\mathrm{d} t + \varphi\right) dt$$

$$= \frac{1}{\pi k_z A^2} \int_{z_0-A}^{z_0+A} \left(F_\Downarrow + F_\Uparrow\right) \frac{z-z_0}{\sqrt{A^2-(z-z_0)^2}} dz, \qquad (2.24)$$

$$I_-(z_0,A) = \frac{2f_\mathrm{d}}{k_z A} \int_0^{1/f_\mathrm{d}} F_\mathrm{ts}\left[z(t),\dot{z}(t)\right] \sin\left(2\pi f_\mathrm{d} t + \varphi\right) dt$$

$$= \frac{1}{\pi k_z A^2} \int_{z_0-A}^{z_0+A} \left(F_\Downarrow - F_\Uparrow\right) dz$$

$$= \frac{1}{\pi k_z A^2} \Delta E(z_0,A) \qquad (2.25)$$

Those integrals have discussed in detail by Dürig[45] and Sader *et al.*[46] Combining Eq. (2.20) and Eq. (2.25), a direct correlation between phase and the energy dissipation is obtained

$$\sin\varphi = -\left(\frac{A}{A_0}\frac{f_\mathrm{d}}{f_0} + \frac{Q_0 \Delta E}{\pi k_z A A_0}\right) \qquad (2.26)$$

Therefore, the phase response of the harmonic oscillator can be used as a qualitative measure for energy dissipation and often reveals chemical contrasts (see Chap. 4.3). The relationship between phase and energy dissipation in Eq. (2.26) can also be found by the conservation of energy principle.[47,48]

Including tip-sample forces, the amplitude and phase response of the cantilever depends on the conservative forces, which are taken into account by the integral $I_+(z_0,A)$, and dissipative forces, which are taken into account by the integral $I_-(z_0,A)$:

$$A = \frac{A_\mathrm{d}}{\sqrt{\left(1 - \frac{f_\mathrm{d}^2}{f_0^2} - I_+(z_0,A)\right)^2 + \left(\frac{1}{Q_0}\frac{f_\mathrm{d}}{f_0} + I_-(z_0,A)\right)^2}} \tag{2.27}$$

$$\tan\varphi = \frac{1}{Q_0} \frac{\frac{f_\mathrm{d}}{f_0} + I_-(z_0,A)}{1 - \frac{f_\mathrm{d}^2}{f_0^2} - I_+(z_0,A)} \tag{2.28}$$

Depending on the sample, the dominant forces between tip and sample can be conservative, dissipative or an interplay between both. Also the quality factor Q_0 as well as the dominant tip-sample forces influence whether the amplitude or phase is the most sensitive signal for the topography feedback loop.

In dynamic mode operation, amplitude modulation AFM is the most successful imaging mode in ambient conditions. In amplitude modulation, the driving frequency f_d is set to a fixed value next to the resonance frequency of the cantilever and also the driving amplitude A_d is kept constant. While approaching the tip towards the sample, the resonance frequency shifts due to distortion of the harmonic oscillator potential by the tip-sample forces. Because of the constant driving frequency, the oscillation amplitude of the cantilever decreases. During scanning, amplitude and phase are measured with a lock-in amplifier while the amplitude is used as the feedback signal for the topography loop. The phase signal is recorded as a dissipation signal. In principle, frequency and phase can also be used for the feedback.

If the phase signal is the more sensitive signal (see Eq. (2.28)), the phase or the cantilevers resonance frequency can be used as a feedback signal. Using the frequency as a feedback signal can be achieved by the simple principle of self-excitation. In frequency modulation mode (constant excitation), only the driving amplitude A_d is kept constant while the driving frequency is adjusted by an additional self-excitation feedback loop to match the resonance frequency of the cantilever in the tip-sample potential. The self-excitation loop only requires a $90°$-phase shifter to feed back the cantilevers oscillation to the driving signal. In most commercial AFM controller electronics, a feedback loop is used to adjust the driving frequency for a constant phase shift of 90°. As a lock-in amplifier is included for amplitude modulation, this can be used for measuring the phase and, therefore, no additional hardware like the phase shifter is needed. The topography feedback loop is adjusting the sample position in frequency modulation (CE) to match a certain frequency offset from the resonance frequency. As this feedback mechanisms require an additional loop to match excitation frequency and resonance frequency compared to the much simpler amplitude feedback, those techniques are mainly used in complex experiments like UHV-AFM or in cases where the amplitude is not very sensitive (e.g. high Q cantilevers).[49]

2.2. Micro- and Nanopatterning Techniques

Modifying surface chemistry and structure is the key to many applications in micro- and nanotechnology. The fabrication of two-dimensional micro- and nanopattern on various substrates forms the foundation of these surface modifications. Fabrication methods to create two dimensional patterns can be separated in top-down and bottom-up approaches. A definition of top-down and bottom-up approaches in nanoscale devices was given by George Whitesides and Christopher Love in 2001: "Nanofabrication methods can be divided into two categories: top-down methods, which carve out or add aggregates of molecules to a surface, and bottom-up methods, which assemble atoms or molecules into nanostructures."[50] Bottom-up approaches often use self-organization processes on the molecular scale to fabricate characteristic patterns by smart interactions of the used materials. Those approaches often offer a simple, fast and cheap fabrication.[51,52] However, specific arbitrary structural features designed for specific applications are often not possible to fabricate with bottom-up approaches. Top-down approaches on the other hand allow the fabrication of specific and application inspired design patterns.[52,53,54,55,56] Top-down approaches rely on lithographic processes to pattern the desired structures. This requires specialized tools to craft the proposed structures in a serial manner (electron beam lithography, laser lithography of scanning probe based methods) or parallel methods using prefabricated masks (UV-lithography, X-Ray lithography). Top-down approaches increase the number of fabrication steps and required tools which is costly in many cases. Therefore, structures fabricated with top-down approaches are often used as a master structure for replication processes like micro- and nanoimprint technologies.[57]

2.2.1. Polymer Blend Lithography

If there is no need of a specific lateral pattern and a distribution of features sizes is fulfilling the requirements of the desired application, polymer blend lithography can provide a fast and cost-effective bottom-up lithographic method. Compared to other methods, this method benefits from the decomposing of two different polymers dissolved in the same solvent, if the solvent is subducted from the blend. Depending on the substrate, the atmosphere, the polymers, the solvent and its concentration, polymer blends can form a purely lateral pattern on a substrate while they decompose during a spin coating process.

The polymer blend lithography used here, utilizes this phase separation effect of a binary polymer blend. The polymers used in this work were polystyrene (PS) and polymethylmethacrylate (PMMA) dissolved in methylethylketone (MEK) (see Chap. 4.3 and Chap. 5.2.4). In general, most polymers remain phase separated when mixed and form an emulsion which has been described in a thermodynamic approach by Flory in 1942.[58] Flory compared the Gibbs free energy G_A, G_B of the spatially separated phases of two polymers and the Gibbs free energy of their mixture G_{AB}. Two polymers, therefore, will form a mixture, if

$$\Delta G_m = G_{AB} - (G_A + G_B) < 0 \tag{2.29}$$

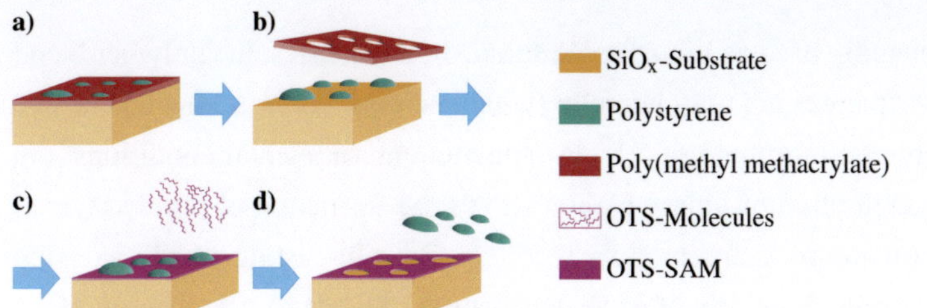

Figure 2.6.: Nanopatterned self-assembled monolayers fabricated by polymer blend lithography. **a)** After a spin-coating process in a controlled atmosphere, the binary polymer blend phase-separates by the temperature drop due to evaporation of the solvent. At a humidity of 45 %, a purely lateral morphology of the phase-separated film is formed with PS islands (green) embedded in a PMMA matrix (red). **b)** Using a selective solvent like acetic acid, a lift-off of the PMMA can be achieved with PS islands remaining on the substrate (by using cyclohexane instead of acetic acid, the PS is lifted off while the PMMA stays on the substrate). **c)** The remaining PS island serve as a mask for the deposition of the OTS-SAM. The OTS molecules are deposited from the gas phase and evaporated at room temperature at 50 mbar. **d)** After the lift-off of the PS islands, the nanopatterned self-assembled monolayer of the OTS-SAM remains on the substrate.

minimizing the free enthalpy ΔG_m for the mixing reaction. As the free enthalpy ΔG_m is also a function of the temperature T, the enthalpy ΔH_m and the entropy ΔS_m of the system $\Delta G_\mathrm{m} = \Delta H_\mathrm{m} - T \cdot \Delta S_\mathrm{m}$, the temperature becomes the critical parameter whether two polymers can coexist in the same phase or remain spatially separated. If a solvent is added, it can be treated as a third quasi-polymer with a polymerization of 1 (just the monomers/molecules of the solvent) which can enable the mixing of all three substances (polymers and solvent) due to a reduction of the Gibbs free energy. For PS and PMMA, MEK can be used as such a solvent. During spin-coating of a substrate with such a solution of a binary polymer blend, the solvent evaporates rapidly and cools the remaining polymer film on the substrate. The temperature drop as well as the Gibbs free energy $G_\mathrm{AB} > 0$ now triggers the decomposing of PS and PMMA, the phase separation. However, a simple phase separation of a binary polymer blend will not necessarily form a pattern, which can be subsequently used as lithographic mask. In most cases, the polymer blend phase-separates not only laterally (which is required for a lift-off) but also in a layered morphology.[59,60] At specific environmental conditions, the phase separation of binary blends allows the formation of purely layered or lateral complex structures.[61,62,63,64,65] By spin-casting on silicon oxide at a moderate humidity of 45 %, the system PS/PMMA dissolved in MEK decays into a purely lateral morphology extending both polymeric phases from the substrate up to the free air interface as described by Huang *et al.*[51] For the lateral separation, the control of the humidity is crucial. For instance in extreme dry atmosphere, the more hydrophobic polymer of the blend wets the air-polymer interface while the more hydrophilic polymer wets the substrate. On the other hand, a very humid atmosphere inverses this trend and additionally water droplets condensate due to the temperature drop while the solvent evaporates forming so-called breath figures.[66]

This technique was used to fabricate masks for the deposition of self-assembled monolayers on SiO_x in a four step process as schematically shown in Fig. 2.6. In the first step, a PS/PMMA mixture in MEK is spin-casted. Following the receipt of Huang *et al.*,[51] the blends morphology is a lateral distribution of PS islands embedded in a PMMA matrix. At this point, the polymer blend lithography is also very flexible in terms of adjusting the structural parameters of the pattern. The size of the PS islands can be adjusted by the molar masses of the PS (smaller molar masses result in smaller islands) and the density by the weight ratio of the dissolved PS and PMMA. In a second step, the desired mask is released using a selective solvent for either to PMMA (acetic acid) or PS (cyclohexane). For the formation of the self-assembled monolayer, organic-trichloro-silanes (OTS) of various chain lengths and 1*H*,1*H*,2*H*,2*H*-perfluorodecyltrichlorosilane (FDTS) were used. The OTS and FDTS molecules are deposited from the gas phase evaporated at room temperature at 50 mbar. In the last step, the remaining mask is removed by dissolving the remaining polymer.

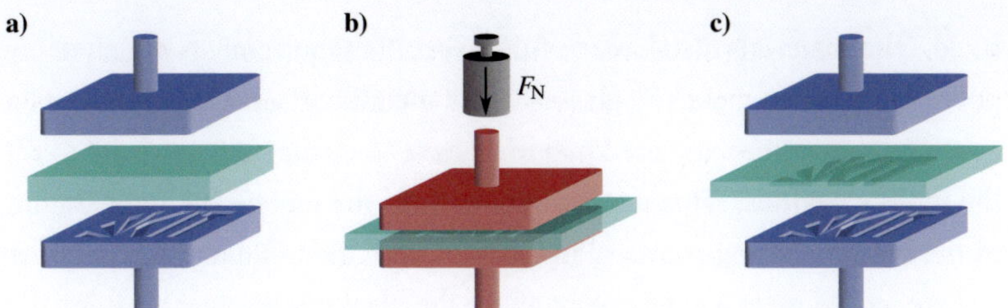

Figure 2.7.: Schematic of the hot embossing process. Hot embossing is a replication process capable of copying a micro and nanostructured master mold into various materials. **a)** In the first step, the mold and the semimanufactured raw material is heated to the embossing temperature. **b)** By controlling position and embossing temperature, the semimanufactured raw material is isothermally deformed to fill the cavities of the mold. **c)** After cooling down to the demolding temperature, the finished sample is demolded. The surface of the sample is a negative copy of the molds surface.

2.2.2. Replication Techniques

If there is a need for a specific lateral pattern and structural details, bottom-up approaches often fail to produce the desired structures. However, top-down approaches have the advantage to fabricate arbitrary lateral patterns and in combination with etching and/or electroplating, they allow the fabrication of 2.5 dimensional structures on various substrates[53,54] with high aspect ratios. Top-down approaches, especially serial writing techniques, are time consuming and use expensive machinery. Therefore, direct lithographically written structures are often disqualified for mass production. One approach to use such structures in a cost-effective and fast way is their replication using a master structure which is replicated several times.

To fabricate a negative copy of the master structure, multiple techniques can be used. In this work, replication into polymer materials has been used. To replicate structures into sheets of raw polymer, hot embossing of microstructures is a flexible method to fabricate even small series of devices.[67,57] The hot embossing process is thereby characterized by three steps (see Fig. 2.7) which have to be optimized for the actual material and mold. The materials also have to fulfill specific requirements which many polymers but also metals,[68] glasses[69] and metallic glasses[70,71] fulfill. At a specific temperature, the used materials have to change the viscosity to fill the mold's cavities. This temperature defines the embossing temperature of the hot embossing process. In the first step, the mold and the polymer foil are heated up to a temperature above the glass temperature allowing the material to flow. In the next step, the material is embossed into the molds cavities. The embossing force pressing both the mold and a substrate plate together, is chosen to fill the molds cavities with the material, to compensate shrinkage and additionally to tune the thickness of the residual material layer between the mold and substrate plate. In the last step, after cooling down the mold and material to a lower demolding temperature, the final device is demolded. The demolding temperature is chosen by the material's properties. The material has to be rigid enough to be demolded and the stress introduced by different heat shrinkage between device and mold material must be kept small enough to prevent any damage to the device or mold before demolding.

While hot embossing is a very general technique to replicate micro- and nanostructures into various materials, some polymers can also be casted into a master mold and cured therein. The raw material thereby are two or more viscous monomers, which are mixed just before casted into the mold. The polymerization of the monomers in the mold form a micro- and nanostructured polymer device which can be demolded after cured.

2.3. Shape Memory Polymers

Certain materials have the ability to remember their initial shape after a plastic deformation. These smart materials can be structured with a permanent shape and can return from a deformed temporary shape to the permanent shape when externally triggered. Shape memory polymers are an especially interesting group of mechanical active materials, because they are capable of single, dual or multiple shape changes[72,73,74] activated by external triggers (e.g. heat or light). Compared with shape memory alloys which allow a displacement in the range of less than 1‰ of the material dimensions,[75] shape memory polymers have a shape changing capability of more than 100%.[76,77] Compared to alloys, they are also price competitive and can be processed like other polymer. However, as the Young's modulus of polymers is in general small compared to alloys, the generated forces for restoring the permanent shape are also small.

In this work, two different commercially available shape memory polymers, Tecoflex® EG 72D and Tecoplast® TP 470, are used. Both allowed the switching from a temporary to a previous structured permanent shape and are thermally activated shape memory polymers which are commercially available by Lubrizol (Ohio, USA). Both are block-copolymers formed by polyaddition reactions of polytetramethylenetherglycol (PTMEG) and 1,4-Butandiol (1,4-BD) with 1-isocyanato-4-[(4-iso cyanatocyclohexyl) methyl] cyclohexan (H12MDI) for the formation of Tecoflex® EG 72D or 1-isocyanato-4-[(4-isocyanato-phenyl) methyl] benzene (MDI) for the formation of Tecoplast® TP 470. They are amorphous, physically cross-linked shape memory polymers[78] with thermoplastic polyether urethanes (TPU) and, as such, linear multiblock copolymers.

The properties of both polymers are defined by a hard segment and a soft segment. As the soft and hard segments also do not mix (as described in Chap. 2.2.1), the shape memory polymer can be described as a phase separated multiblock copolymer with a low temperature phase and a high temperature phase. As the polymers are heated above $T_{g,ht}$ of the high temperature phase, also the low temperature phase is above its glass transition temperature. The polymer can be processed like any thermoplastic polymer, e.g. by hot embossing. While above $T_{g,ht}$, the permanent shape can be defined and fixated by cooling of the polymer below $T_{g,lt}$ (structuring of the permanent shape). The high temperature phase now provides the netpoints of the permanent shape. By heating the polymer above the transition tem-

Tecoflex® EG 72D

Tecoplast® TP 470

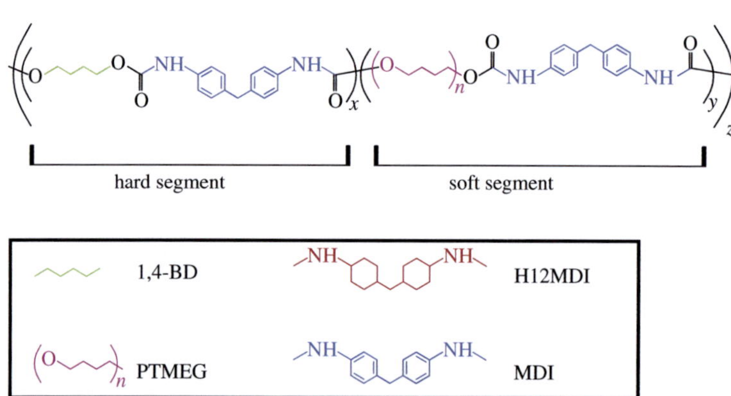

Figure 2.8.: Structural formula of Tecoflex® EG 72D and Tecoplast® TP 470. Both polymers are phase segregated polyurethane multiblock polymers. Both phases have different glass transition temperatures $T_{g,hd}$ and $T_{g,sd}$. Although both polymers have a similar structure Tecoplast® TP 470 has higher transition temperatures than Tecoflex® EG 72D. The transition temperature of the hard segment is tuned by the use of MDI instead of H12MDI which has a higher melting temperature. The soft segment can also be tuned by the degree of polymerization n of the PTMEG which is a polymer formed by ether reaction out of the 1,4-BD.

perature of the low temperature phase, the stabilization of the shape by the soft segments is switched off and the shape can be temporary deformed. As long as the high temperature phase is below its transition temperature, the deformation of the hard domains and therefore of the netpoints of the permanent shape is elastic. If cooled below $T_{g,lt}$ while deformed to a temporary

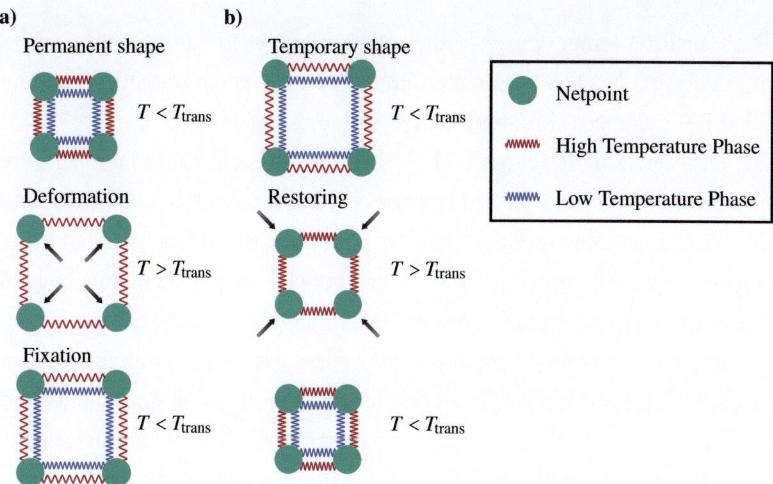

Figure 2.9.: Schematic of programming shape memory polymers. The structure is defined by the interplay between the low and high temperature phases stabilizing the shape. The stabilization by the low temperature phase can be switched on and off with temperature. **a)** In the permanent shape, low and high temperature are stabilizing the shape. While heated above T_{trans}, the low temperature phase stops stabilizing the shape. The high temperature phase can now be deformed elastically. To stabilize the temporary shape, the polymer is cooled down and the low temperature phase stabilizes the temporary shape. **b)** To restore the permanent shape, the polymer is heated above T_{trans} to suppress the stabilization of the temporary shape by the low temperature phase. The permanent shape is restored by a relaxation of the high temperature phase.

shape, the low temperature phase stabilizes the temporary shape (fixation of the temporary shape). The shape memory effect allows a switching back to the permanent shape by heating the polymer above $T_{g,lt}$. The switching temperature is therefore called shape transition temperature T_{trans}. Without external forces deforming the material and internal forces stabilizing the temporary shape, the elastic deformation of the high temperature phase is reversed by a relaxation of the material. This cycle of deformation of the permanent shape, fixation of the temporary shape and restoring is shown in Fig. 2.9.

The transition temperatures of the segments are defined by its monomers. Pure PTMEG has a very low melting temperature of around 30°C, adding H12MDI enhances this temperature to around $50\,°C - 75\,°C$.[78,77,79,80,81] The hard segment of Tecoflex® EG 72D has a glass transition temperature of around $120\,°C - 155\,°C$ and is formed by adding 1,4-BD to H12MDI. As this transition temperature is significantly higher, it is the netpoint-forming hard segment. Additionally, the hydrogen bonds between the double bonded oxygen and the hydrogen bonded to the nitrogen stabilize the intramolecular formation of netpoints. The variance of glass transition temperatures can be explained by variations in the molecular weight of the monomers and mixing ratios of the segments. As those parameters like the parameters n, x, y and z (see structural formulas in Fig. 2.8) are business secrets of Lubrizol, effective values for $T_{g,ht}$ and the switching temperature T_{trans} have to be found experimentally.

For Tecoplast® TP 470, similar arguments apply. However, the H12MDI of Tecoflex® EG 72D has been replaced by MDI which has higher melting temperatures. Also the mixing ratios are assumed to be different from Tecoflex® EG 72D. Tecoplast® TP 470 therefore shows higher temperatures both for $T_{g,ht}$ and T_{trans}. The temperatures of the polymers used here are $T_{trans,TFX} = 55\,°C$, $T_{g,ht,TFX} = 150\,°C$ and $T_{trans,TP} = 85\,°C$, $T_{g,ht,TP} = 195\,°C$.

3. Scanning Probe Microscopy Utilizing Cantilevers With Integrated Sensors

Since the invention of the atomic force microscope in the 1980s,[4] it has become a versatile tool used in nanoscale metrology, biosensing, maskless lithography and high density data storage with nearly as many sensing techniques as applications.[55,82,83,84] In general, the force measurement with a AFM is done by deflection measurement of a micro-machined cantilever. For deflection sensing, two different detection systems are competing. Sensing with an external deflection sensor and self-sensing devices. Most state of the art instruments used in ambient conditions rely on an optical read-out of a micro-fabricated cantilever,[21,85] while for specific applications and environments like vacuum self-sensing tuning forks with manually attached tips are desirable.[86,87,88] Micro-machined cantilevers can be mass fabricated but the optical read-out contains bulky mechanical parts to focus a laser on the backside of the cantilever and to move the position sensitive photo-detector or a mirror. While adjusting the laser and photo-detector is easy in ambient conditions where all components are accessible, it is a challenge in other environments like vacuum or in fluids where the laser gets scattered and refracted by multiple interfaces.[89,90] Furthermore, optical read-outs have to be readjusted not only after every cantilever exchange but also after drifts in the environmental conditions like temperature because it can offset the focal position of the laser and photo-detector due to thermal expansion. Additionally, the optical read-out can also influence the cantilevers deflection, as it is utilized for photothermal excitation of the cantilever[6] and interfere with the sample as it can cause photobleaching of fluorescence samples.[7] On the other hand, self-sensing tuning forks suffer from the limited capability of mass-production and a reduced number of operational modes compared to micro-machined silicon or silicon-nitride cantilevers. Additionally, cantilevers fabricated by silicon based microfabrication methods allow the integration of multiple additional features like doping for better electrical conductance or the integration of active sensing elements. Magnetoresistive sensors for example can be used for strain and deflection sensing[8] and thermoresistive sensors enable nanoscale thermal analysis.[9, 91]

Therefore, an ideal AFM sensor should allow various operational modes, is self-sensing and has the capability of mass production. Previous attempts to tackle this problem mostly used integrated piezoresistive and piezoelectric sensors on micro-machined cantilevers to measure either the strain[92, 93, 94, 95, 96, 97, 98] in the lever or its displacement.[99, 100] One part of this work focuses on magnetoresistive strain sensors for self-sensing AFM cantilevers in order to sens tip-sample forces. While magnetoresitive strain sensing can be used to simplify instrumentation, an other part of this work focuses on sensing specific tip-sample interactions. Thermoresistive sensors integrated to AFM cantilevers next to the tip enable a highly sensitive detection of heat flux along the tip into the sample. Therefore, scanning probe microscopes equipped with such cantilevers are called scanning thermal microscopes[11] and represent a new class of instruments for nanoscale thermal analysis. This chapter focuses on sensor concepts, implementation and details of instrumentation.

3.1. Setup of a Nonmagnetic Large-Area-AFM

In order to characterize magnetoresistive strain sensors integrated into AFM cantilever, the cantilever deflection must be measured in parallel by independent means. This requires an instrument with an external deflection sensor for micro-machined cantilevers. However, commercially available instruments with a beam deflection setup are not suitable for those specific

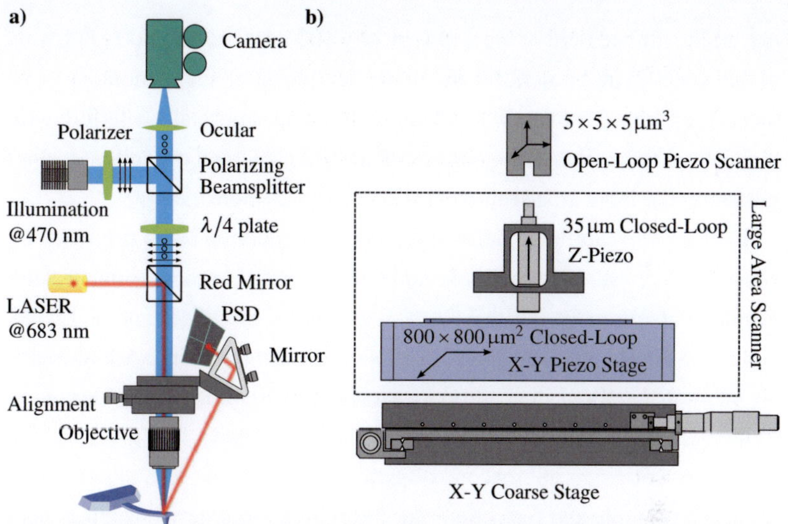

Figure 3.1.: a) Optical setup of the beam deflection read-out on the large area AFM. The laser is focused with an optical microscope objective on the backside of the cantilever. The reflected beam is focused on the PSD with a tilting mirror to align the beam towards the PSD. The illumination of the sample is done at a different wavelength than the laser wavelength to decouple laser and illumination by selective mirrors in the microscope and filters at the PSD. To reduce stray light within the optical microscope, polarizing optics for the illumination allow a directed transmittance of the microscope image. **b)** To realize both high lateral resolution and a large field of view, a high resolution open loop scanner is nested on a large area closed loop scanner. For coarse positioning of the sample, the entire scanner can be move with respect to the cantilever and optics.

cantilevers as this work aims on the investigation of the influence of magnetic fields to the magnetoresistive sensors. Most commercially available instruments show intrinsic magnetic properties and can distort applied magnetic fields. Therefore, the setup of a custom build nonmagnetic scanning probe microscope is required. As the instrument must be equipped with an external deflection sensor, also conventional AFM cantilevers without integrated sensing can be used in this instrument. This offers the chance to extend the limited field of view of typically $100 \times 100\,\mu m^2$ to $800 \times 800\,\mu m^2$ at this specific instrument for the characterization of shape memory polymers (see Chap. 6 for details). As large scan ranges normally interfere with high lateral resolution, a new concept of a small area, high resolution scanner nested on the large area scanner is successfully demonstrated.

To sense the cantilever deflection, a beam deflection setup proposed by Meyer *et al.*[21] was also realized and is implemented in most commercially available instruments. For this study, it allows the use of commercial AFM controller electronics and control software implementations. Additionally, an integrated optical microscope for coarse navigation on the sample is implemented. A schematic of the optical setup is shown in Fig. 3.1. Using an infinity corrected microscope objective and an ocular lens allow to illuminate the sample and focus the laser beam on the cantilever with the same objective. Using the microscope objective to focus the laser also simplifies the adjustment of the laser beam deflection setup because the complete optical microscope can be moved instead of adjusting the laser. As a result, the focal spot of the laser is fixed towards the field of view of the optical microscope and the laser is aligned to the cantilever when the cantilever is at a specific position in the optical image. To block scattered light inside the optical path of the laser from the camera, a red mirror is used to couple the laser beam into the objective. As the mirror reflects only light with wavelengths longer than 600 nm, all light from the laser is either reflected towards the objective or the laser itself. The cantilever is tilted toward the optical axis of the microscope and acts as a mirror for the laser beam. As the cantilever gets deflected, the angle of the cantilever tilts towards the incident laser beam and consequently the reflection angle changes. As the reflected beam is divergent (due to the focusing of the microscope objective), it is refocused to the position sensitive photo-detector by a tilting mirror.

To illuminate the sample, a wavelength shorter than the reflection edge of the red mirror was chosen. To suppress stray light within the optical path of the microscope, it is useful to use polarizing optics. In contrast to the laser,

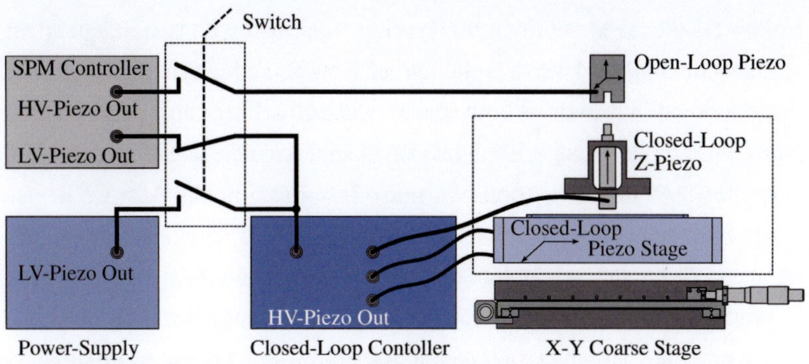

Figure 3.2.: Principle of nested high resolution scanner on the large area scanner. The system consists out of two independent scanning units: the large area closed loop scanner and a nested open loop high resolution scanner with a small scan area. The SPM controller can provide both, a high voltage signal for the small area scanner and a low voltage signal for the closed loop controller of the large area scanner. To switch from large area scanning to high resolution scanning, the large area scanner can be moved and hold on the desired sample position due to its independent closed loop controller while the SPM controller continues scanning with the open loop scanner.

the light of the illumination has first to illuminate the sample, gets reflected at the sample and the reflected light has to pass the complete microscope to the camera. By using polarized light for illumination, a polarizing beamsplitter can be used to reflect all light from the light source of the illumination towards the sample. By passing a $\lambda/4$ plate, the polarization direction gets rotated by $45°$. After being reflected on the sample, the light passes the $\lambda/4$ plate again and the polarization is rotated again by $45°$. The polarization of the reflected light is now $90°$ rotated towards the incident light from the light source. Therefore, the beamsplitter is completely transparent to light reflected from the sample which can pass towards the camera.

The AFM is operated with a commercial AFM controller (ARC2-biPolar by Asylum Research). The controller can directly drive open-loop piezo scanners, because of its integrated high voltage amplifier, as well as closed-loop scanners with an attached high voltage amplifier and closed-loop controller. As the AFM set up in this work is equipped with two independent scanners to combine both, a large field of view and a high spatial resolution, this feature allows to drive both scanners directly. Large area scanners normally show a higher noise level than small area scanners due to lever motion amplifiers. As the elongation of a piezo is approximately $\Delta L = \pm E \cdot d \cdot L_0$, where E is the applied electric field, d the piezoelectric coefficient of the material and L_0 the initial length of the piezo with typical values for piezo stack actuators of $U = \pm 220\,\text{V}$, $d = 350\,\text{pm/V}$ and a distance between two electrodes of 1 mm. To achieve a travel of $800\,\mu\text{m}$ by direct drive, approximately 1 m of piezo ceramic per axis is required. Therefore, motion amplification using levers is a suitable way to reach such travels. However, levers used as motion amplifiers also have some drawbacks. Assuming an ideal lever with a lever transmission ratio r, a travel of the actor ΔL_{act}, a stiffness of the actor k_{act} and a resonance frequency of the actor f_{act}, the corresponding characteristics of the amplified scanner can be calculated from:

$$k_{sys} = \frac{k_{act}}{r^2} \tag{3.1}$$

$$\Delta L_{sys} = \Delta L_{act} \cdot r \tag{3.2}$$

$$f_{sys} = \frac{f_{act}}{r} \tag{3.3}$$

By using levers to amplify travel, the stiffness and dynamic of the scanner are reduced. The reduced resonance frequency increases the response time of the scanner to driving signals. Therefore, lever amplification can only used for the slow lateral scanning as the z-axis of the scanner needs a high resonance frequency for high dynamics. The large area scanner has a motion amplified x-y piezo stage and a dedicated z-piezo for high dynamics. Additionally, the x-y stage must only move in the x-y plane without any cross-talk to the z-axis. This is reached by flexure joints. A flexure is a frictionsless, stictionless device based on elastic deformation (flexing) of

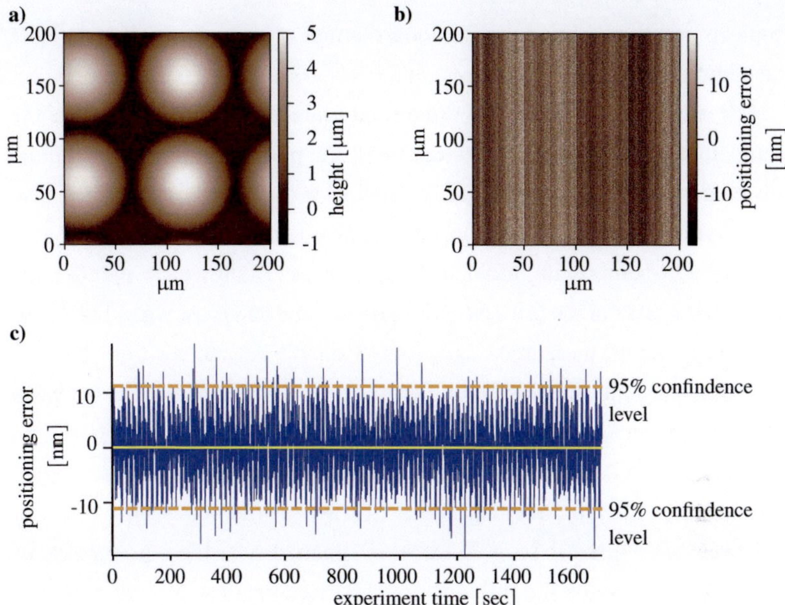

Figure 3.3.: A crucial precondition for a nested high resolution scanner design is the stability of the underlying large area scanner. The position accuracy and positioning error can be tested by reading out the sensing elements for the closed loop system. **a)** A $200 \times 200\,\mu\text{m}^2$ topography image of microlenses. While scanning the microlenses, the read-out of the closed loop sensor in the fast scan direction is recorded. **b)** The read-out of the fast scan axis is compared with the desired scan position and a positioning error can be extracted. The origin of vertical lines in the picture can be assigned to a switching of the closed loop controller between different sensitivity ranges. **c)** To demonstrate the stability of the large area scanner, the positioning error is constant below 10 nm for typical imaging times of more than 30 minutes. The sensor is read-out with a sampling rate of 1.5 kHz.

a solid material. However, as the stiffness of a lever amplified system k_{sys} is reduced quite significantly, the initial stiffness of the flexure stage has to be quite high. As a result, no scanner which fulfill those requirements was commercially available and had to be developed by Physik Intrumente (PI) in Karlsruhe specifically for this application. All other components of the scanner like the capacitive positioning sensors of the closed-loop system, z-piezo and the additional open-loop scanner are commercially available standard components.

For successful switching to the nested scanner, first the stability of the large area scanner has to be tested. First, the positioning accuracy can be tested during AFM scanning. If scanned with the open-loop scanner, also the stability and drift of the large area scanner is of interest. In Fig. 3.3 a), a scan of a hot embossing mold with microlenses is shown. In parallel, the positioning error of the fast scan axis was recorded and is shown in Fig. 3.3 b). By comparing the measured stage position and the desired position (given by the control signal), the positioning error was extracted. While the feedback loop of the stage is fully analog to avoid errors from any analog-to-digital converter (ADC), the measurement electronics to record the sensor signal are not. Specifically, one nanometer of travel corresponds only 6 tics of a 16-bit ADC if the maximum travel is spanned over the whole range of the ADC. To reduce the influence of the digitalization, a preamplifier is used to amplify and offset the signal suitable for the ADC input. However, the switching between these different sensitivity ranges causes some errors visible in the error signal as vertical stripes. On the other hand, not only positioning accuracy but also long time stability of the scanner is of interest. In Fig. 3.3 c) the positioning error for typical experiment times of up to 30 minutes is measured. The data shows now drift of the stage during the whole experiment and only small fluctuations around the desired position by ±10 nm which is a remarkable low value for a scan stage which has a maximum travel of 800 μm.

As the large area scanner is very stable, it can be used to carry a second small area scanner with a higher spatial resolution and better dynamic properties. Using an AFM with multiple scanners allows both, a large field of view and a high spatial resolution. In Fig. 3.4, the potential of such an instrument is demonstrated. By scanning a calibration grating structure with structural details on length scales spanning from hundreds of micrometer to less then 200 nm and a feature height of 22 nm, the topography of the

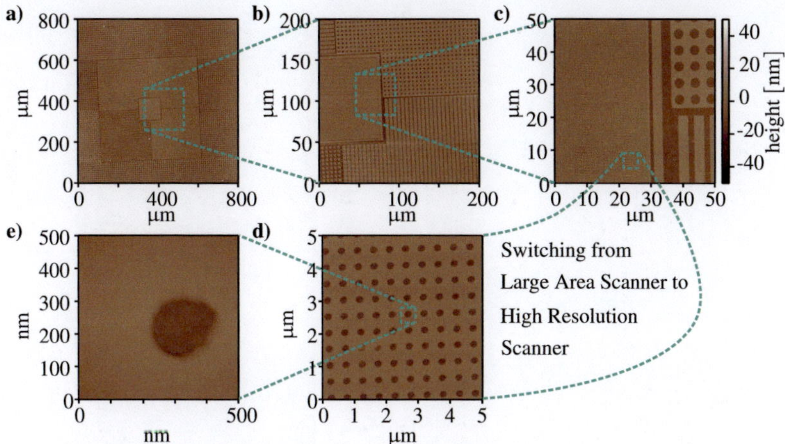

Figure 3.4.: Large area scanning with switching to the small area scanner for high resolution. All 5 pictures where taken on a SiO_x calibration grating with various feature sizes. **a)** An $800 \times 800\,\mu m^2$ overview image of the structure was taken before the scansize was reduced to $200 \times 200\,\mu m^2$ in **b)**. In **c)** the scansize was reduced again to $50 \times 50\,\mu m^2$. After switching to the high resolution scanner **d)** and **e)** show $5 \times 5\,\mu m^2$ and $500 \times 500\,nm^2$ images of the smallest feature sizes of the calibration grating. The nested scanner which can span over 3 orders of magnitude in scan range makes this instrument a versatile tool for micro and nanomechanical analysis.

sample can be investigated on all length scales. For a first overview of the sample, the maximum scan size can be used and sequentially zoomed into the region of interest. As the desired zoom level results in a scan size below the maximum scan range of the high resolution scanner, the scan position can be hold with the large area scanner while the sample is scanned with the small area scanner enabling further zoom steps. Thereby, the instrument can span over 3 orders of magnitude in scan range which makes it a unique tool for micro- and nanomechanical analysis.

One example of such analysis is given in Fig. 3.5. For quality control of fabrication steps in microstructure technology, the AFM is often used for spot checks of the fabricated structures. However, as most AFMs are limited to a field of view of $100 \times 100\,\mu m^2$, they are only suitable for local imaging.

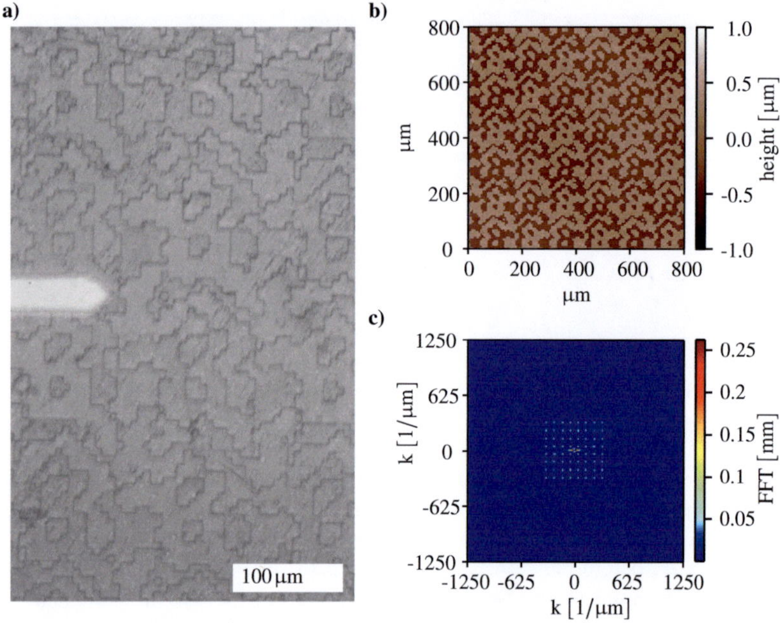

Figure 3.5.: Quality control of a mold insert of a optical phase grating. **a)** The optical microscope image allows a fast overview of the structure. **b)** The AFM topography can be used to control the homogeneity of the structure height on scales larger than the grating period of $256 \times 256\,\mu m^2$. **c)** The quality of periodic grating structures however can be tested in the k-space revealing by using an FFT on the topography image.

Often, features of structural details will just not fit into this field of view. An example of such structures are periodic structures like optical phase gratings fabricated at the Institute of Microstructure Technology.[101] Imaging such structures with the large area scanner allows to image a whole grating period of $256 \times 256\,\mu m^2$ in a single picture. Such diffractive optical structures define the length of the optical path of light propagating through, by their topography. The interference pattern of such structure can be described by the Fourier transformation of the topography. By Fourier transforming the AFM topography image, the interference pattern of diffractive structures is revealed and can be used for quality inspections of the mold.

3.2. Magnetoresistive Strain Sensors

As pointed out above, an ideal AFM sensor should allow various operational modes, is self-sensing and has the capability of mass production. Previous attempts using integrated piezoresistive and piezoelectric sensors on micromachined cantilevers sens strain or displacement of the cantilever. However, in principle those sensors show sufficient sensitivity for the implementation in atomic force microscopes but suffer from reliability in mass fabrication.[98] Driven by the increasing demand for magnetic hard disk drives,[102] magnetic tunneling junctions[103, 104, 105, 106, 107, 108, 109, 110] are state of the art read-heads in magnetic hard drives. Additionally, they can be adapted for high strain sensitivity[111] and offer remarkable miniaturization opportunities.[112] In combination with already implemented processes of mass fabrication, they are a promising alternative to piezoresistive and piezoelectric sensors for self-sensing AFM cantilevers.

3.2.1. Magnetoresistive Tunneling Junctions

For sensing local magnetic fields, the discovery of the anisotropic magnetoresistive effect (AMR) in 1856[113] enable the technical foundation of modern hard disk drives. However, more than one century later the discovery of the giant magnetoresistive effect (GMR) in 1986[114] (Nobel price in Physics 2007 to Albert Fert and Peter Grünberg), the development of aluminum oxide tunnel junction (MTJs)[115, 116] and magnesium oxide tunneling junctions[117, 118, 110] (Millenium Technology Price 2014 to Stuart Parkin) increase the sensitivity and, therefore, the storage density on magnetic disks. In Fig. 3.6, a schematic of a magnetic tunnel junction is given. The magnetic tunnel junction thereby consist out of two ferromagnetic electrodes separated by a thin dielectric layer. As the thickness of the dielectric layer is ranging from a few angstroms to a few nanometers, electrons can tunnel through the barrier. It is, therefore, often referred as the tunneling barrier. The tunneling of electrons through a dielectric layer arouses an electrical current and is fully quantum mechanical. It can be explained by the wave nature of the electrons while the resulting junction conductance depends on the evanescent state of the electron wave function within the tunneling barrier. The general tunnel current between two electrodes of the same material is given by Simmons' expression from 1963:[119]

$$I(U) = \varphi(t_b)\left(\left(\bar{\Phi} - \frac{U}{2}\right)e^{\left(-1.025\sqrt{\bar{\Phi}-\frac{U}{2}}\,t_b\right)} - \left(\bar{\Phi} + \frac{U}{2}\right)e^{\left(-1.025\sqrt{\bar{\Phi}+\frac{U}{2}}\,t_b\right)}\right)$$

(3.4)

Simmons expressed in this equation the tunneling current I as a function of the barrier height $\bar{\Phi}$, the applied bias voltage U across the junction and its thickness t_b.

In magnetic tunneling junctions, both electrodes consist out of a ferromagnetic material, in which the electric current is split into two partial currents. Those partial currents are conducted in two sub-bands each carrying either spin-up or spin-down electrons. The sub-bands are spin polarized. During tunneling through the barrier, the electron spin must be conserved (see Fig. 3.6) and the tunneling current depends on the relative magnetic orientation of the electrodes towards each other. The electrons split up into spin-up $N_\uparrow$ and spin-down $N_\downarrow$ electrons and as a result they can only tunnel between sub-bands with the same spin orientation. For a parallel magnetic orientation, both electrodes show the same sub-band configuration. The tunneling occurs between two electrodes with the same density of states (DOS) at the Fermi level E_f. In an antiparallel magnetic orientation, however, a mismatch between the density of states occurs, as the sub-bands of one electrode are exchanged. Consequently, the tunneling current is decreased in this case. However, in actual tunneling barrier the electron spin can be influenced by magnetic impurities like impurity atoms within the barrier. To conserve the electron spin during tunneling, the barrier has to be extremely pure. The conductance of a magnetic tunneling junction, therefore, strongly depends on the orientation of the magnetization of the electrodes towards

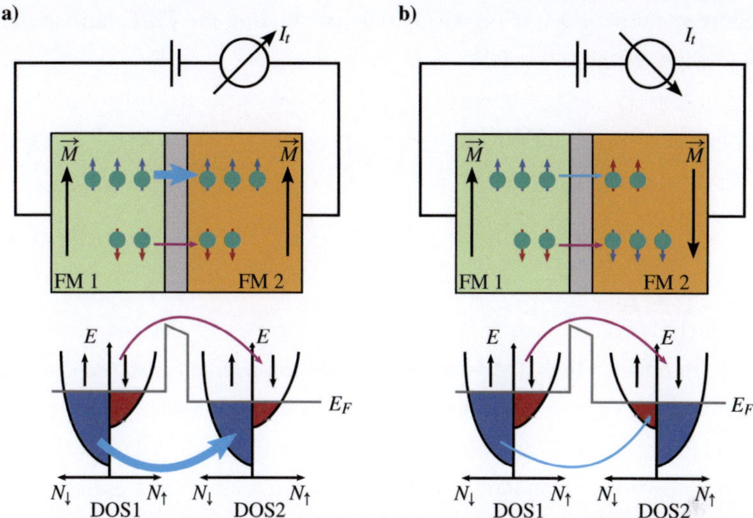

Figure 3.6.: Principle of magnetoresistive tunneling junctions. During tunneling, the electron spin orientation is preserved and electrons can only tunnel between sub-bands with the spin orientation. The conductance of the magnetic tunneling junction is therefore proportional to the product of density of state at the Fermi level of the spin polarized sub-bands of the two electrodes. Changing the orientation of the two electrodes from a parallel state **(a)** to an anti-parallel state **(b)** inverts the sub-band orientation of the second electrode and causes a change in the conductance.

each other. It can be expressed by the angle α between the magnetization of the two electrodes:

$$G(\alpha) = \frac{1}{2}\left(G_p + G_{ap}\right) + \frac{1}{2}\left(G_p - G_{ap}\right)\cos\alpha \tag{3.5}$$

$$TMR_{ratio} = \frac{G_p - G_{ap}}{G_{ap}} = \frac{R_{ap} - R_p}{R_p} \tag{3.6}$$

Here, G_p, R_p, G_{ap} and R_{ap} are the conductance and resistance for the parallel ($\alpha = 0°$) and antiparallel ($\alpha = 180°$) state. A measure for the resistance change between the parallel and antiparallel state is the so-called TMR ratio.

Julliere specified Eq. (3.6) in 1975[115] by correlating the TMR ratio to the polarization of the sub-bands:

$$TMR_{ratio} = \frac{2P_1 P_2}{1 - P_1 P_2} \tag{3.7}$$

$$P = \frac{N_\uparrow (E_F) - N_\downarrow (E_F)}{N_\uparrow (E_F) + N_\downarrow (E_F)} \tag{3.8}$$

with the polarization factors P_1 and P_2[120, 121] for the two electrodes, respectively. To obtain a high TMR ratio and, therefore, a large resistance change, electrodes with a high spin polarization factor are preferable. With a polarization factor $P = 0.61$, CoFeB electrodes have a high spin polarization and can be grown with standard sputtering techniques[108, 122] and used for the fabrication of the magnetic tunneling junctions used in this work.

As the model of Julliere is based on the assumption of amorphous tunneling barriers fabricated from AlO_x, it only describes incoherent tunneling through the barrier. Crystalline MgO barriers like the ones used in this work allow coherent tunneling. Therefore, the model of Julliere does not apply to those barriers. However, for crystalline barriers, the coherent tunneling can be described analytically using classical solid-state physics assumptions.[117] The coherent tunneling current can be calculated using the Landauer[123] conductance which relates the conductance to the probability of a Bloch electron transmitted from one electrode through the MgO barrier to the other. By modeling the electrodes as two electron reservoirs with the chemical potential $\mu_{1,2}$ connected with the tunneling barrier, Butler *et al.*[117] expressed the tunneling current

$$I = I^\rightarrow - I^\leftarrow = \frac{e^2}{h} \sum_{k_\parallel, j} T^\rightarrow (k_\parallel, j) \frac{\mu_1 - \mu_2}{e} \tag{3.9}$$

with the transmission probability $T^\rightarrow (k_\parallel, j)$ (left electrode to right electrode) for all currents j for all Bloch states for a given value of $k_\parallel$ by the individual currents $I^{\rightarrow,\leftarrow}$ from the left electrode to the right electrode and counter-wise. Equation (3.9) transforms the equations which describe plane waves being transmitted through the barrier into a set of equations for Bloch waves. In the case of ideal coherent tunneling, the Bloch states of the electrodes $\Delta_1 (spd)$, $\Delta_5 (pd)$ and $\Delta_2 (d)$ couple with the evanescent states (tunneling states) in the band gab of the MgO barrier $\Delta_1 (spd)$, $\Delta_5 (pd)$ and $\Delta_{2'} (d)$, respectively.[117,118,124] As all conduction channels show different decay lengths, this can greatly enhance the TMR ratio. While the $\Delta_1 (spd)$ channel has the longest decay length, the $\Delta_{2'} (d)$ has the shortest decay length. The $\Delta_1 (spd)$ channel is therefore dominant. For the parallel state, the $\Delta_1 (spd)$ Bloch state of the majority band of one electrode couples via the tunneling barrier with the $\Delta_1 (spd)$ Bloch state of the other electrode. The tunneling discussed for this parallel state implies the same symmetry for the band-structure of both electrodes. For the antiparallel state however, a combination of the features of the majority and minority channels is observed[117] because of the broken symmetry. While the Bloch states from the majority band of one electrode, e.g. the $\Delta_1 (spd)$, can still couple with the tunneling states of the barrier, theses states can not propagate because there are no minority $\Delta_1 (spd)$ states at the Fermi energy in the second electrode and are reflected by the second electrode. As a result, only the rapidly decaying states in the tunneling barrier can propagate, reducing the tunneling current significantly.

3.2.2. Magnetostrictive Electrodes

Magnetic tunneling junctions like described above are excellent magnetic field sensors and, therefore, often used in read-heads of magnetic hard drives.[122] Using magnetostrictive materials in the electrodes of the magnetic tunneling junction,[125] the magnetization of one electrode can rotate if strained. For the classical applications of magnetic tunneling junctions, this is an unwanted effect as it can alter the field measurement.[126]

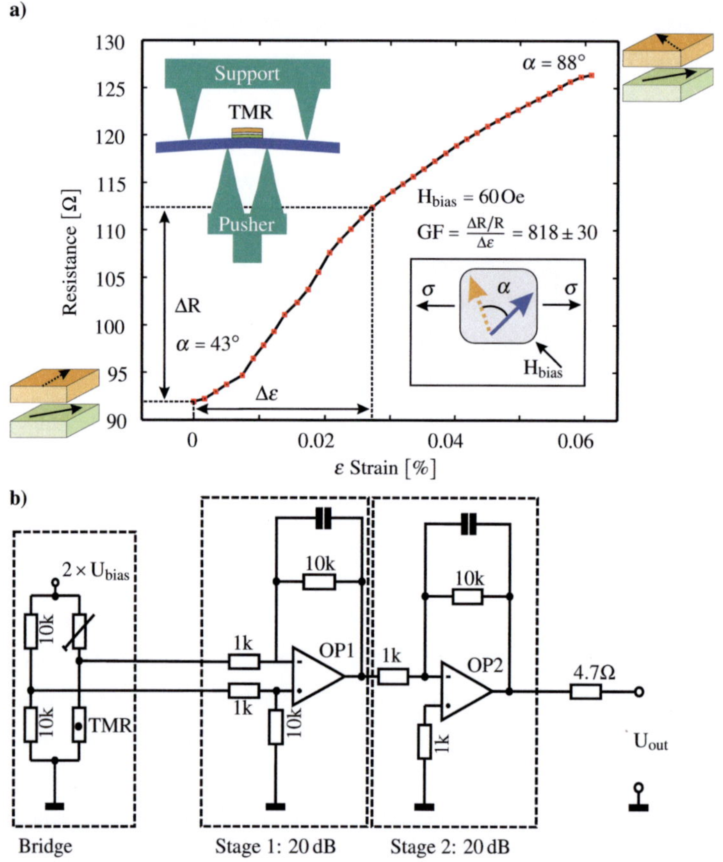

Figure 3.7.: Resistance of TMR sensors with magnetostrictive sensing electrodes as a function of applied strain. **a)** Using the inverse magnetostrictive effect, the magnetization of an electrode rotates when strain is applied. This results in a change of tunneling current in the magnetic tunneling junction. Using a 4-point bending apparatus, the resistance of a $10\,\mu m \times 10\,\mu m$ sized TMR sensor is measured as a function of the applied strain revealing a high gauge factor of over 800. Reproduced with permission from A. Tavassolizadeh.[8] **b)** To apply magnetostrictive tunneling junctions to an AFM cantilever, the resistance change must be measured with high precision at a low bias voltage at the TMR sensor of 10 mV. This requires a low noise, high gain and bandwidth amplifier with the TMR sensor in a Wheatstone bridge configuration. To fulfill these requirements, the design has to be compact to be integrated as close as possible to the TMR sensor.

To describe this effect, an effective field H_{eff} exerting a torque on the magnetization in the Landau-Lifshitz-Gilbert Eq. (3.10), which describes a magnetic moment under the influence of a magnetic field, must be discussed[127]

$$\frac{\partial M}{\partial t} = -\frac{\gamma}{1+\beta^2} M \times H_{eff} - \frac{\beta}{(1+\beta^2)M} M \times \left(M \times H_{eff} \right) \qquad (3.10)$$

with the gyromagnetic ratio γ and the Gilbert damping parameter β. To obtain H_{eff}, the total magnetic energy E_{total} of an electrode can be described by[128]

$$E_{\text{total}} = E_{Exchange} + E_{Uniaxial} + E_{Zeeman} + E_{Demag} + E_{magnetoelastic} \qquad (3.11)$$

$$H_{eff} = -\frac{1}{\mu_0} \frac{\partial E_{\text{total}}}{\partial M} \qquad (3.12)$$

in an equilibrium magnetization configuration. To change the magnetization of one electrode, the total magnetic energy has to be modified. To discuss the influence of E_{total} on H_{eff}, the individual terms of E_{total} can be discussed. The exchange energy $E_{Exchange}$ has to be constant, because it describes the overlap of adjacent atoms in the electrode.[129] The uniaxial energy is the crystalline anisotropy due to the crystallographic directions of the electrode. The magnetization of a crystalline electrode prefers to align along certain crystalline axes which are often referred as the easy axes.[130] As the atomic and crystalline structure will not change, also the corresponding energies can be assumed as a constant. The Zeeman energy is the potential energy of a magnetization configuration in an externeal field. As the external field is kept constant during the measurement, also the Zeeman energy is a constant. The magnetostatic energy refers to domain wall formation within the ferromagnetic electrode which can reduce the total magnetic energy. Therefore, a single domain electrode is assured during the measurements by a saturation of the electrode along the easy axes with a high bias field before measurements. The magnetoelastic energy can be changed by applying strain due to the inverse magnetostrictive effect. The magnetoelastic energy has been found to be highly strain dependent[131]

$$E_{magnetoelastic} = -\frac{3}{2}\lambda_{iso}\sigma\cos^2(\Theta)$$ (3.13)

$$\sigma = \varepsilon E$$ (3.14)

with the isotropic saturation magnetostriction λ_{iso}, the stress σ, strain ε, Young's modulus E and the angle Θ between the magnetization M and the direction of the stress. As the magnetoelastic energy is the only energy which can be varied by strain, this energy is the only energy influencing the effective field H_{eff}.

Because of the inverse magnetostrictive effect,[130] the magnetization of the electrodes rotates when they are strained. To use this effect for strain sensing, only one electrode must rotate when strain is applied to the junction. Therefore, the magnetic tunneling junction has to be included into a TMR stack, which includes contact electrodes and a pinning mechanism to fix the magnetization of one electrode. One of the two electrodes is used as a reference layer while the second electrode is free to rotate. To fix the magnetization of the so called reference layer, it is pinned to an artificial antiferromagnet. The exchange bias between the antiferromagnet and the reference layer fixates the magnetization of the reference layer. If the reference layer has a fixed magnetization, the resistance of the tunneling junction can be tuned by rotation the magnetization of the free sensing layer. Using the inverse magnetostrictive effect in the sensing layer makes the TMR stack sensitive to applied strain. The stack was grown on silicon wafer level by Karsten Rott and Günter Reiss at Bielefeld University. Subsequently, the wafer was processed into AFM cantilevers with highly sensitive TMR strain sensors by Ali Tavassolizadeh at Christian-Albrechts-University zu Kiel. In Fig. 3.7 a), the resistance to strain response of such a sensor on the wafer is given. Using such TMR cantilevers in an AFM can thereby greatly simplify the design of an instrument as shown in Fig. 3.8. During the design of an AFM with a laser beam deflection setup, most effort has to be spend on designing the optical read-out. Using self-sensing cantilevers greatly reduce the effort. A good measure is the amount of parts, which has to be machined for the AFM with the optical read-out and for the AFM, which can be operated with self-sensing cantilevers only. In the instrument presented earlier in this chapter, the optical read-out only is assembled out of 48 specifically machined parts. However, to obtain a high TMR ratio, a low bias voltage is needed. Zhang and White[132] pointed out in 1998, that the tunneling current is strongly temperature and bias voltage dependent. They propose a two step tunneling via defect states in the tunneling barrier which is spin independent. These defect states thereby are governed by the Fermi-Dirac function

$$f(E) = \frac{1}{1 + \exp\left[(E_c - E)/k_B T_{eff}\right]}, \tag{3.15}$$

with the energy level of the defect state E, the energy level of the barrier conduction band edge E_c, the Boltzmann constant k_B and the effective barrier temperature T_{eff}. The effective temperature is found as a function of both, the temperature T and bias voltage U:

$$k_B T_{eff} = \sqrt{(C k_B T)^2 + (e V_1 \exp(U/V_2))^2} \tag{3.16}$$

with the free parameters V_1, V_2 and C representing material properties of the barrier.

Therefore, small bias voltages of 10 mV are used in this work. Additionally, even small voltages of a few millivolt generate high electric fields on the nanometer thick barrier increasing the risk of discharge and destroying the junction. However, this leads to small tunneling currents which are hard to measure. In contrast to the measurement show in Fig. 3.7 a) which is done with a long integration time constant and low pass filtering, this type of measurement is not suitable for an application on AFM cantilevers. The read-out of magnetic tunneling junction read-heads in hard drives is done very fast. However, in a hard drive, there are only two different states of magnetization. In the AFM, not only two different states must be observed, but also all states in between the parallel and antiparallel state. This can be achieved by using the TMR sensor in a Wheatstone bridge configuration with an attached high gain and bandwidth amplifier. As the bridge voltage using a bias voltage of only 10 mV will only vary by a few nanovolts for small deflections of the cantilever and, therefore, small resistance variations, the amplifier also has to show a low intrinsic noise level. This requires as less components as possible (as every component adds intrinsic noise) and short wires between TMR sensor, bridge and amplifier as they act as antennas picking up RF-noise (see Fig. 3.7 b)). As a result, the cantilever is mounted on a printed circuit board and connected via wire-bonds to the integrated amplifier. The low bias voltage can potentially be increased in further experiments to increase the signal-to-noise ratio even as this will reduce the TMR ratio to optimize the electric working point of the sensor.

3.3. Thermoresistive Sensors

To measure thermal transport on the nanoscale, experimental methods are certainly at their infancy. Scanning thermal microscopy,[11] which is adapted from atomic force microscopy is a promising microscopy technique to investigate thermal transport on extremely small scales as it is the only experimental technique with nanometer-scale spatial resolution in thermal imaging. The key element in scanning thermal microscopy is a sharp tip probing the heat flux during scanning over a sample. As scanning thermal microscopy is similar to atomic force microscopy, models describing the contact mechanics can also be used for thermal analysis. Compared to atomic force microscopy cantilevers, a temperature sensor is integrated into the cantilever next to the tip. The cantilevers temperature is measured by measuring the temperature dependent electrical resistance of the integrated

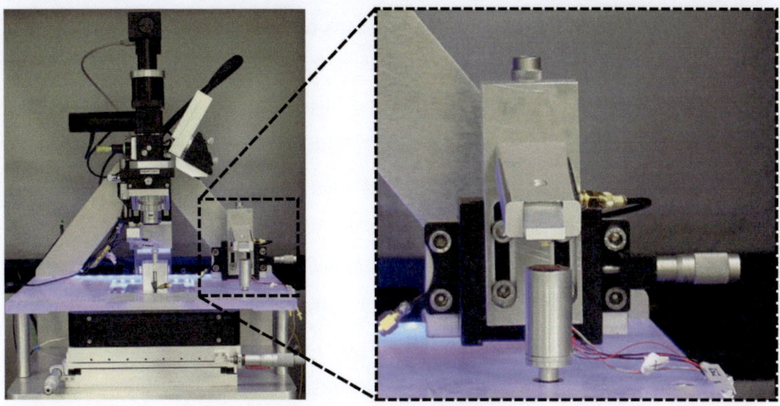

Figure 3.8.: Comparison of the large area AFM with optical read-out and a TMR sensor read-out. Using cantilevers with integrated sensing elements can greatly simplify the instrumentation design. On the left, a photograph of the AFM with the large area scanner and the optical read-out is shown. Next to it and in the zoom on the right, a compact design using the $5 \times 5 \times 5\,\mu m^3$ open loop scanner is show. This instrument has no optical read-out and fully relies on cantilevers with TMR sensors.

sensor. While the cantilever is heated with a constant power, its temperature is measured. This technique is sensitive to heat flux variations due to thermal conductance variations between tip and sample. Because of the small dimensions of the thermal probe, its heat capacity is also very small and it can, therefore, change its temperature very fast. Its time constant to thermally equilibriate the tip and sensor, which limits the bandwidth of the scanning thermal microscope, is approximately $50\,kHz$[9] and sensitive to the interface between tip and sample.[91]

In more detail, a thermoresistive sensor is integrated in the cantilever on the base of its tip and not only used as a sensing element. It is also used to heat up the tip with a constant power to bias the thermal measurements. The probes are home-built at IBM Zurich Research Laboratory and micro-fabricated out of mono-crystalline silicon[10] with sharp tips of typical 500 - 700 nm height (schematically shown in Fig. 3.9 a)). The cantilevers are highly phosphorus doped (10^{20} at/cm^3) for good electric conductivity and a lower nominal doping density of 5×10^{17} at/cm^3 in a $4\,\mu$m $\times\, 6\,\mu$m $\times\, 200$ nm large volume at the tips base forming the heater/sensor with a temperature dependent electrical resistance of few kΩ. The tips used in the measurements presented in Chap. 5 are typically conical with a spherical tip apex for a good thermal conductance of the tip and a spring constant of around 0.3 N/m. A scanning electron microscopy micrograph of a tip used in the experiments of Chap. 5.2.5 is shown in Fig. 3.9 c). Quantitative thermal measurements are complicated, as energy in the form of heat can be transported by all materials, all states of matter and even radiation. Quantitative experiments must, therefore, been carried out in vacuum to suppress heat flux from the heated region to the sample via surrounding air. At vacuum conditions, the heat is only conducted along 3 heat paths: the tip-sample-path, along the cantilever and via radiation. The heat transfer via radiation can be neglected, because it is at least 3 orders of magnitude smaller than the tip-sample conductances.[133] The overall thermal conductance of the cantilever in vacuum can be measured to be around $4.3\,\mu$W K^{-1}, while the tip is not in contact with a sample.

3.3.1. Temperature Calibration

For a quantitative analysis of the measured heat flux, the heater/sensor is calibrated by measuring its electrical resistance as a function of the temperature by its current-voltage response ($I - V$ curve) of the heater/sensor[9, 12, 134] (see Fig. 3.9 d)). Assuming that all electrical power is dissipated in the heater/sensor, the increase of power dissipated results in an increase of the electrical resistance due to the increased scattering of carriers in the heater/sensor (ohmic response of the thermoresistor). With rising temperature, the number of thermally generated carriers in the semiconductor rises over the number of dopants and decreases the electrical resistance for higher temperatures. The temperature at the maximum electrical resistance is a

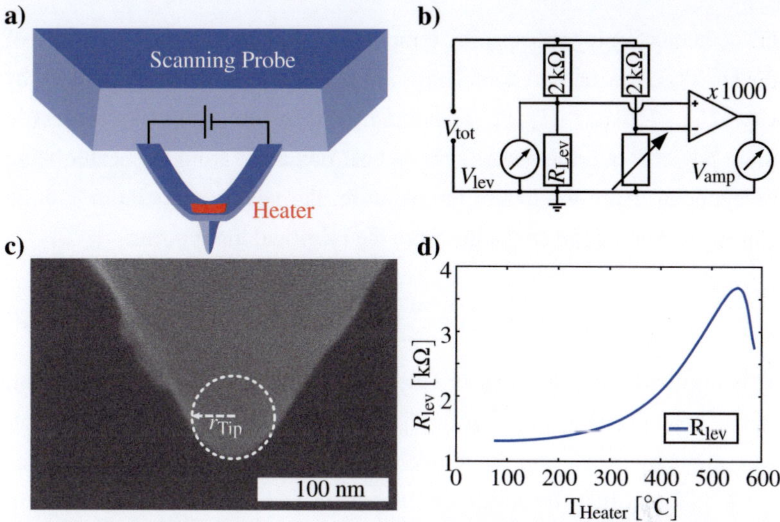

Figure 3.9.: a) Schematic of cantilever and heater/sensor. During the experiment, the low doped heater/sensor (red) is electrically heated with constant electrical power. **b)** Circuit for the heater/sensor read-out. The electrical power dissipated in the heater/sensor is determined by the current (via the voltage drop at the series resistor of $2\,k\Omega$) and the precise measurement of the voltage drop over the lever by combining V_{lev} and V_{amp}. **c)** SEM micrograph of a tip used for thermal measurements. The micrograph was recorded after several experiments and shows no sign of contamination or traces of blunting at the apex. A relatively blunt spherical tip is in good agreement to the assumptions in the applied contact model. (Image courtesy: Bernd Gotsmann, IBM-Zurich) **d)** Heater/sensor resistance as a function of the temperature. By measuring the current voltage response ($I-V$-curve) of the heater/sensor, its temperature can be determined by the temperature-dependent electrical resistance of the heater/sensor.

function of the doping density[135] and occurs for the heater/sensor used in this work at $T_{Rmax} = 550\,°C$. This characteristic maximum resistance for the given doping density can be used as calibration reference. As the $I-V$-curve is measured with the tip not in contact with a sample and at vacuum conditions, the temperature T_{heater} for the maximum electrical resistance is given by $T_{heater} = T_{Rmax} = 550\,°C$ by knowing the doping level of the heater. Using the specific temperature T_{Rmax} and the temperature of the lab T_{Lab}, the thermal conductance of the cantilever can be calculated by $G_{th} = P_{Rmax}/(T_{Rmax} - T_{Lab})$. Assuming the heat flux $\dot{Q}$ equals the electric input power P into the heater and the heat flux away from the heater being independent from the heater temperature, the heater temperature can be expressed using a linear scaling from the electrical input power:

$$T_{heater} = T_{Lab} + \frac{(T_{Rmax} - T_{Lab})}{P_{Rmax}} \cdot P = T_{Lab} + \frac{1}{G_{th}} P \qquad (3.17)$$

This implicitly assumes that the thermal conductance G_{th} is independent of the heater/sensor temperature T_{heater}. When checking all of these assumptions by measuring the temperature by independent means, Menges *et al.*[12] found that the resulting systematic error is below measurement errors and fabrication tolerances given a relative temperature resolution (sensitivity) of $\approx 0.5\,\mathrm{mK}$ with an absolute uncertainty (accuracy) of about 30 %. For a precise measurement of the electrical power dissipated in the heater, a high precision series resistor is used while the voltage drop over the heater/sensor is determined by a combination of a direct measurement across the heater/sensor and an amplified read-out of a balanced Wheatstone bridge configuration (see Fig. 3.9 b)). The read-out is quite similar to the read-out of the TMR-Sensors described in Chap. 3.2.2, however, one major difference has to be pointed out. For quantitative analysis, the absolute electrical resistance of the sensing element has to be measured very precisely. For the TMR-Sensors, a relative resistance change was sufficient. Therefore, the amplified voltage of the Wheatstone bridge was monitored for precise measurements of variations. Additionally the resistance was measured directly. Combining both, the direct measurement with the precise relative measurement, results in a high precision measurement of the absolute resistance.

A shortened version of the first part of this chapter is in preparation as the article "A Scanning Probe Microscope for Magnetoresistive Cantilevers Utilizing a Nested Scanner Design for Large Area Scans", T. Meier, A. Förste, A. Tavasollizadeh, K. Rott, G. Reiss, E. Quandt, D. Meyners, R. Gröger, T. Schimmel and H. Hölscher: Review of Scientific Instruments (2014).

4. Magnetostrictive Sensing

Even though atomic force microscopy is a versatile tool for modern nanotechnology,[82,55,83,84] still most commercial instruments rely on external deflection sensors to sens the deflection of mass fabricated micro machined cantilevers. Optical deflection sensors are not only bulky compared to instruments relying on self sensing cantilevers (see Fig. 3.8), they can also influence the cantilever deflection[6] and interfere with the sample and other characterization methods.[7] To avoid photo bleaching and background fluorescence, instruments specialized for biological samples often use lasers in the infrared regime for a beam deflection setup. However, using wavelengths invisible to the eye requires the use of additional equipment for alignment. The concept of self-sensing cantilevers based on magnetic tunneling junctions as it is presented in this chapter can help to overcome these issues as they show sufficient sensitivity for use in atomic force microscopes. The application of magnetostrictive materials as strain sensors has been proven by Mamin *et al.*[136,137] while magnetoresistive displacement sensing was employed by Sahoo *et al.*[138] to atomic force microscopy.

Here, the integration of self-sensing cantilevers based on magnetostrictive tunnel magnetoresistance sensors (TMR) to atomic force microscopy is introduced since they offer high strain sensitivity[111] and remarkable miniaturizing opportunities.[112] This makes them a promising alternative to already employed piezoresistive[92,94,95,96,97,98,139] and piezoelectric sensors.[93,98] The magnetostrictive TMR sensor consists out of two ferromagnetic layers, the so-called reference and sensing layer, separated by an insulating tunneling barrier. The reference layer is magnetically pinned to an artificial antiferromagnet sandwiched by the exchange bias effect to a natural antiferromagnet.

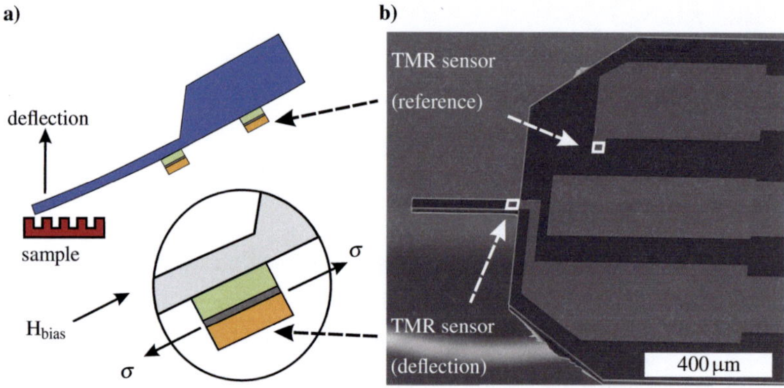

Figure 4.1.: a) A schematic of a self-sensing microcantilever based on magnetostrictive TMR sensors. Upward deflection causes tensile stress σ on the bottom side of the cantilever. This is detected by a resistance change of the deflection TMR sensor. The TMR sensor on the cantilever chip is not subjected to stress and can be used as a reference. **b)** A scanning electron microscopy image (bottom view) of a self-sensing microcantilever including two TMR sensors on the cantilever and its chip. Reproduced with permission from A. Tavassolizadeh.[8]

The magnetostrictive sensing layer on the other hand is free to rotate. As the TMR effect is a spin-dependent electron tunneling phenomenon, the tunnel resistance depends on the angle ϕ between the magnetizations of these two ferromagnetic layers. Due to the inverse magnetostrictive effect, induced strain rotates the easy axis of the positive magnetostrictive sensing layer and varies thereby the tunnel resistance.

4.1. Integration of Magnetic Tunneling Junctions to Microfabricated Cantilevers

The fabrication of the cantilevers with TMR sensors was done by Ali Tavassolizadeh and Dirk Meyners at Christian-Albrechts-University in Kiel, Germany, while the deposition of the TMR stack was done by Karsten Rott and Günther Reiss at Bielefeld University, Germany, within a joint DFG project. The incorporation of TMR sensors into AFM cantilevers is thereby conceived over a sequence of MEMS techniques in order to ensure durability of TMR sensors properties within the fabrication process. The TMR-stack is grown by sputtering techniques on a $4''$ Si(100) wafer substrate with $300 \pm 2\,\mu$m thickness (Si-Mat Silicon Materials, Germany). It is covered with $2\,\mu$m-thick and $100\,$nm-thick silicon dioxide, which is thermally grown on the back and front side, respectively. The process starts with transferring windows into the backside silicon dioxide layer for later anisotropic etching of silicon by means of photolithography and reactive ion etching (RIE). On the front side of the wafer, the TMR stack is deposited using a sputtering system with a base pressure of 2×10^{-7} mbar.

The TMR stack thereby not only consist out of two ferromagnetic layers and a tunneling barrier but also out of electric contacts, an antiferromagnet providing the exchange bias and interface layers. The layers of the stack are: Ta (5 nm) / Ru (30 nm) / Ta (10 nm) / Ru (5 nm) / MnIr (12 nm) / CoFe (3 nm) / Ru (0.9 nm) / CoFeB (3 nm) / MgO (1.8 nm) / CoFeB (3 nm) / Ta (5 nm) / Ru (5 nm). The reference layer of this stack is the MnIr (12 nm) / CoFe (3 nm) / Ru (0.9 nm) / CoFeB (3 nm) section. While the MnIr serves as a antiferromagnet coupled to the CoFe, the coupling to the CoFeB electrode can be tuned by the thickness of the Ru spacer. Furthermore, an annealing step is carried out at about $360\,^\circ$C for 1 h at a pressure of 10^{-6} mbar under a magnetic field of $2\,$kOe. This procedure leads to crystallization of the CoFeB layers and to smooth CoFeB/MgO interfaces.[140] It also aligns the easy axis of the sensing layer and pins the reference layer due to the imposed magnetic exchange bias.[141] As the TMR sensor must be sensitive to both tensile and compressive stress, the reference layer is aligned in a 45° angle towards the cantilever axis. In a next step, TMR sensors with sizes between $10\,\mu$m $\times$ $10\,\mu$m to $37\,\mu$m $\times$ $37\,\mu$m are defined together with the required contact pads by photolithography, ion beam etching, and lift-off techniques. Back-side alignment is performed to pattern the TMR sensors at the fixed end and the support chip of the future cantilevers. As depicted in Fig. 4.1

the first TMR sensor experiences maximal strain during bending while the second sensor is fixed and only incorporated as a reference sensor for optional differential measurements.

TMR and RA values of the final sensors measured in magnetic field loops under a 10 mV bias voltage were 120 % to 220 % and 14 kΩµm^2 to 61 kΩµm^2, respectively. The fabrication proceeds on the back side with silicon anisotropic etching up to 280 µm in depth through the initially opened windows by 40 % KOH at 80 °C. Finally, cantilevers are released by another photolithography and RIE process applied on the front side. The cantilevers used in this study were 300 µm to 700 µm long, 20 µm to 50 µm wide, and 10 µm to 20 µm thick. The cantilevers are designed relatively thick to avoid pre-bending and pre-strain after the complete fabrication process.

4.2. Strain Sensitivity and Contact Mode AFM Imaging

As depicted in Fig. 4.1 a) this detection principle of a magnetostrictive TMR sensor can be easily applied to measure the bending of a AFM cantilever. The alignment of the initial easy axis of the sensing layer is set to 45° against the applied stress. In this way the TMR sensor is sensitive to both compressive and tensile stress which is required for essentially all modes of AFM.[142] In particular, TMR sensors with CoFeB/MgO/CoFeB magnetic tunnel junction are well known for their very high TMR values.[143] In addition, the use of a $Co_{40}Fe_{40}B_{20}$ sensing layer leads to high strain sensitivity as demonstrated in Fig. 3.7 a). Those measurements, however, are done with a 4-point bending apparatus and a magnetic bias field of 60 Oe perpendicular to the magnetization of the pinned reference layer and with tensile stress applied to the junction. On the cantilever level, not only tensile but also compressive stress occurs. Assuming single domain behavior of the

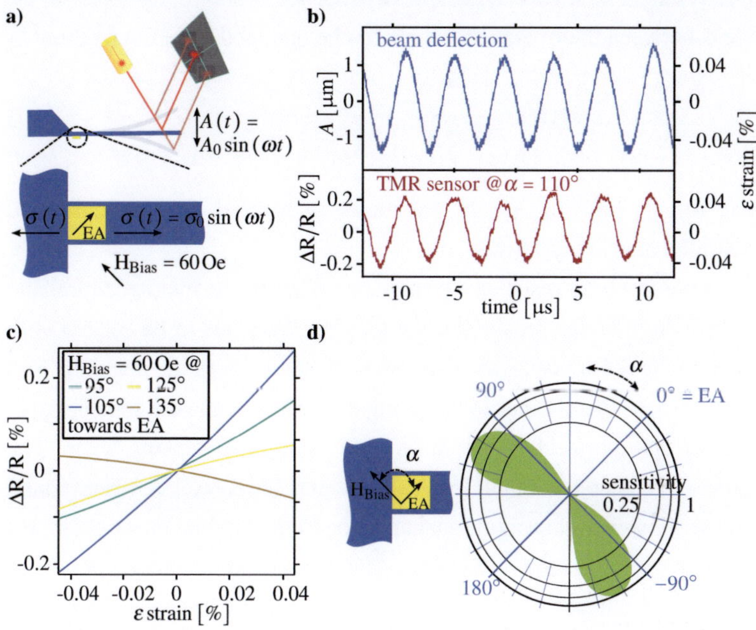

Figure 4.2.: Analysis of the sensitivity of a TMR sensor as a function of the direction of the bias field. **a)** The cantilever is oscillated at its resonance frequency while the amplitude is monitored with the beam deflection setup. In parallel, the resistance of the TMR sensor is measured while the direction of the bias field is varied. **b)** The oscillation of the cantilever is quantified by the beam deflection setup and causes strain in the cantilever at the TMR sensor position. In this notation, tensile stress corresponds to positive strain. The resistance change of the 27 μm × 27 μm sized TMR sensor can be correlated to the applied strain. **c)** The resistance change as a function of strain is exemplary plotted for four different angles of the bias field. The bias field has a strong influence on the strain sensitivity of the TMR sensor as well as the direction of the applied stress. **d)** The normalized sensitivity as a function of the incident angle of the bias field. As the easy axis is oriented 45° towards the cantilever, the easy axis is also aligned 45° toward the applied stress. The sensitivity response is symmetrical with respect to the easy axis (EA).

two ferromagnetic layers and cosine dependence of the conductance G, the angle ϕ between their magnetizations can be obtained from the relation[144]

$$R(\phi) = \frac{1}{G(\phi)} = \frac{R_\perp}{1 + \frac{R_{ap}-R_p}{R_{ap}+R_p}\cos(\phi)},\qquad(4.1)$$

where R_p, $R_\perp$ and R_{ap} are the resistance in parallel, perpendicular and antiparallel state, respectively.

To achieve a high resistance change per angle of the magnetization and thereby a high strain sensitivity, the sensing layer has to be rotated with respect to the reference layer. This can be achieved with the magnetic bias field. The field thereby has to be strong enough to rotate the sensing layer but weak enough to allow strain induced rotation of sensing layer. The optimum of the field strength for this particular TMR stack has been found at 60 Oe. However, the angular dependence of the applied bias field is still an open question. As the setup of the AFM allow both, the measurement of the cantilever deflection by independent means and the response of the TMR sensor as a function of the angle of the magnetic bias field, the field can be varied while the optimum is found. The resistance of the 27 µm × 27 µm sized TMR sensor changes under the applied tensile stress induced by oscillating the cantilever at its resonance frequency (see Fig. 4.2 a) and b)). The deflection of the cantilever is directly measured with the beam deflection setup of the AFM. With the deflection, the strain at the base of the cantilever can be approximated using Hooke's law $E = \frac{\sigma}{\varepsilon}$ with the applied stress and the Young's modulus:

$$\varepsilon = \frac{\sigma}{E} = \frac{6lk}{ba^2E}\Delta z\qquad(4.2)$$

$$k = \frac{Eba^3}{4l^3}\qquad(4.3)$$

using the the length l, width b, height a and the displacement Δz of the rectangular cantilever. In Fig. 4.2 c) the sensor response for four chosen field angles is given. The strain sensitivity (slope of the sensor response) varies quite significantly with the incident angle of the magnetic field. The sensor also shows a higher sensitivity for tensile strain which can be used in pre-strained junctions or to distinguish between compressive and tensile stress for spectroscopy applications. The normalized sensitivity as a function of the angle is given in the logarithmic polar plot of Fig. 4.2 d). The TMR junctions with a squared geometry used in this work show the highest strain sensitivity at a bias field angle α of $115°$ towards the initial magnetization of the reference layer. The angle α is therby defined as the angle between the easy axis and the bias field in the range of $0°$ to $180°$. The angle of the bias field was varied in $5°$ steps while the TMR sensor was saturated along the easy axis between each angle variation. For symmetry reasons, the behavior of the TMR sensor can be assumed to have the same sensitivity for negative values of α, however, the signal from the TMR sensor is inverted with respect to the signal for positive values of α.

Such sensors are indeed more suitable for AFM purpose than the giant magnetoresistive sensor with a gauge factor of 150 used by Mamin *et al.*[136] as the TMR sensors used here show a gauge factor of up to 820 (see Fig. 3.7 a)). In order to use the cantilevers with integrated TMR sensors as self-sensing cantilevers, contact mode AFM was tested first. The measurement of the relative resistance change of the TMR sensor was realized by its integration into a Wheatstone bridge configuration with 20 mV bias voltage. The voltage drop on the TMR sensor is in the unstrained configuration at 10 mV and the bridge is kept balanced. The voltage between the midpoints was amplified by 60 dB and low-pass filtered with a cut-off frequency of 300 kHz. This read-out of the TMR sensor was directly fed into an analog-digital converter of the AFM controller. For contact mode measurements shown in the following, a bias field of 60 Oe was applied perpendicular to the magnetization of the reference layer and the read out of the TMR sensor

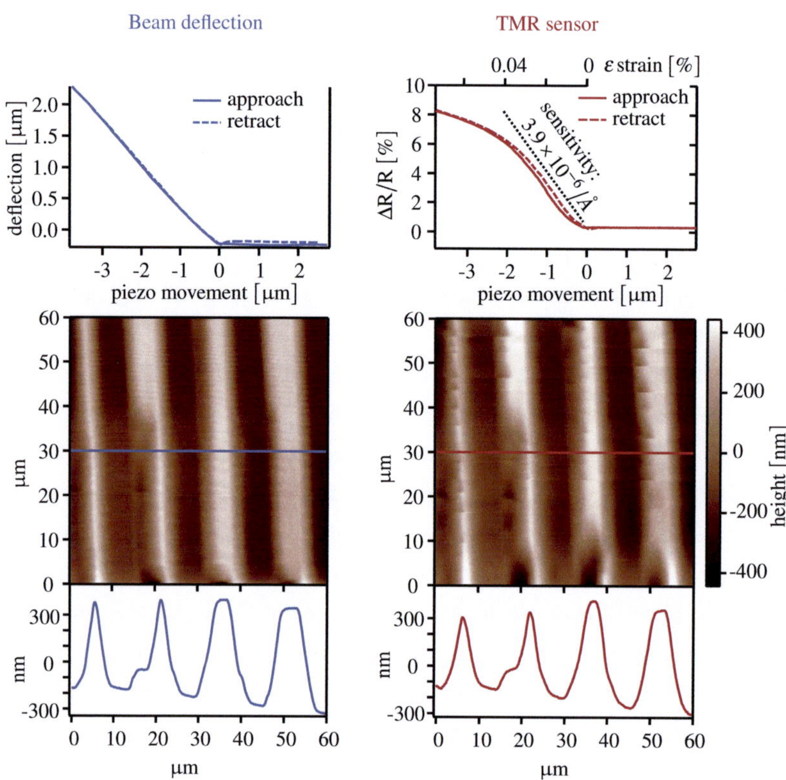

Figure 4.3.: Comparison of force vs. distance curves simultaneously measured with the optical beam deflection system (left) and the TMR sensor (right). The hysteresis between the approach and retraction curves for the TMR sensor is probably caused by magnetic hysteresis effects in the sensing electrode. The dashed line in the TMR curve represents a sensitivity of 3.9×10^{-6}/Å. The axis on top shows nominal strain acting on the TMR sensor. The topography images and line sections of a PMMA line grating are recorded with a tipless cantilever subsequently on the same sample position with the feedback on the beam deflection and the TMR sensor. After the first scan with the feedback on the beam deflection, the image recorded with the TMR sensor reveals some damages on the sample caused by drag forces during the first scan. As the PMMA sample is relatively soft, the topography was modified during the first scan and the subsequent image (with the feedback on the TMR sensor) is different as it reveals kinks at the walls of the grating.

was recorded in parallel to the conventional optical beam-deflection signal. In this way, one of both signals can be arbitrarily chosen for the feedback during scanning of the sample. The bias field angle is chosen different from the angle for the maximum strain sensitivity. In this way, the sensing layer is not aligned for maximum strain sensitivity in the unstrained configuration but for a strained configuration which is an advantage in contact mode.

The bending of the cantilever was measured during its approach and retraction towards the sample surface (force vs. distance curve). Figure 4.3 displays the simultaneous read-out of the TMR sensor and the beam deflection system. A comparably large travel of the z-piezo (about 6 μm) was chosen in order to monitor the resistance change also for high strains in the TMR sensor. The strain response from the TMR sensor shows a clear non-linear response to the applied strain which could possibly be related to a magnetic saturation in the direction of the stressanisotropy. In this particular case, the highest sensitivity of the TMR sensor can be observed for a strain of 0.02 %. The minor hysteresis observed for the TMR sensor is probably caused by magnetic hysteresis of the sensing electrode. The deflection signal is roughly linear for the optical beam-deflection while a non-linear relation to the applied z-displacement of the TMR sensor ($\Delta R/R$) is revealed. However, the deflection of the cantilever during those force vs. distance curves is large compared to the typical deflection below 100 nm during most AFM experiments. On this deflection ranges, a roughly linear sensor response can be assumed.

From the force vs. distance curves in Fig. 4.3, the deflection sensitivity (relative change in resistance divided by the cantilever deflection) can be calculated. The maximum value obtained from the displayed data is about $3.9 \times 10^{-6}/\text{Å}$ (see dashed line in Fig. 4.3). This value is already better than previously reported values for metallic based[145] and semiconductor based[146, 147, 148, 149] piezoresistive sensors which are specifically optimized as strain sensors in AFM cantilevers.

Scanning in contact mode with a self-sensing TMR cantilever is presented in Fig. 4.3 on a PMMA grating. As the first cantilevers manufactured for this project had no integrated tips, one of the cantilever apexes was used for scanning.[150] However, as the cantilevers width is around 50 µm, high drag forces between cantilever and sample occur. To compare the image quality in contact mode, the sample was first scanned with the feedback on the laser beam deflection setup. In a second scan on the same sample position, the feedback was switched to the TMR sensor. As the PMMA sample is relatively soft, the topography was damaged during the first scan (with the feedback on the laser beam deflection) and the subsequent image (with the feedback on the TMR sensor) is different as it is revealing kinks at the walls of the grating. Both images show the same imaging quality for the feedback on either the TMR sensor and the laser beam deflection.

4.3. Dynamic Mode Imaging

To reduce drag forces, the AFM was switched to amplitude modulation. Still using the same sample system, Fig. 4.4 shows a comparison of the results obtained in dynamic mode (amplitude modulation) with the TMR and optical read-out.

In this case, the scanning in the dynamic mode is not effecting the sample due to the absence of friction forces. The resonance curves (amplitude and phase) shown on the top of the Fig. 4.4 are nearly the same for both detection principles. Since the oscillation amplitudes commonly used in amplitude modulation mode are considerably smaller as the piezo movement applied in the force-vs.-displacement curves presented in Fig. 4.3, the nonlinearity of the TMR sensor is negligible in this mode. The oscillation amplitude of both the beam deflection read-out and the TMR sensor are given in volts as the raw signal from the detector. The oscillation of the cantilever (in nanometer) is in both cases the same as both curves are recorded in parallel. The different absolute scales in the amplitude is, therefore, related to different

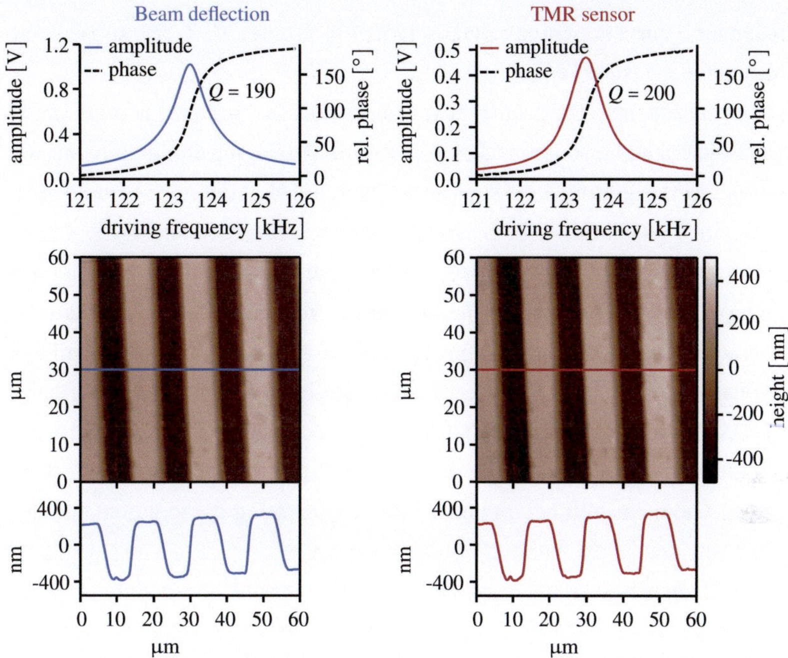

Figure 4.4.: Amplitude modulation mode imaging with beam deflection (left) and TMR sensor (right). On top are the resonance curves simultaneously measured by both read-outs. The topography images of an optical grating are displayed at the bottom and were subsequently recorded at the same sample position using the conventional beam deflection and the TMR sensor signal as feedback, respectively. Line sections taken from the same positions are shown at the bottom of the two AFM images and reveal that the data quality obtained with both sensors is the same. As no drag forces occur in amplitude modulation mode, the grating structure is not damaged like in the previous contact mode imaging.

gains in the PSD read-out electronics and the TMR read-out electronics. Consequently, the quality factors calculated from the two amplitude curves are of the same order ($Q_{\text{Laser}} = 190$ and $Q_{\text{TMR}} = 200$). The difference in the Q-factors might thereby be related to the different strain sensitivity for tensile and compressive stress like shown in Fig. 4.2, but more likely, the difference is caused by the scattering of subsequent measurements of the resonance curves. This scatter is typically around 10 % for subsequent resonance measurements.

During scanning the cantilever was mechanically oscillated at its mechanical fundamental resonance ($f_d = 123.5\,\text{kHz}$). Again, the topography images of the PMMA grating were subsequently recorded at the same sample position with a feedback on the conventional beam deflection and TMR sensor, respectively. The absence of friction forces enables a stable imaging over multiple scans without damaging the sample. Additionally, the "tip"-sample interaction is now confined on one apex of the cantilever beam which allows to image finer structure details. The data quality observed in the topography of the PMMA grating imaged with the feedback on the TMR sensor in amplitude modulation mode is the same than the quality of the beam deflection setup. The measured height and width of the grating are identical and the same tiny details can be found in both images.

For an increased lateral resolution, however, the contact geometry between cantilever and tip has to be improved. Using electron beam deposition, tips can manually been grown on the apex of the cantilever.[151] The use of manually grown tips allows high lateral resolution as tip radii of down to 30 nm have been achieved here. As an additional advantage, the fabrication process of the TMR cantilevers must not be modified as the tip growth can be done afterwards. The growth of tips has been done by Dirk Meyners at Christian-Albrechts-University Kiel.

Using self-sensing cantilevers with attached tips, the sensitivity of the cantilevers can be investigated further. Using the small scan-size open-loop scanner, features on smaller scales can be investigated. The ultimate challenge in terms of sensitivity using strain sensing cantilevers is the imaging of atomic step edges. As setups with a beam deflection system can routinely image such small features, only the ability of the TMR sensor to reveal such features is of interest. Figure 4.5 shows such atomic step edges on gold(111) terraces. This image was obtained in amplitude modulation with the feedback on the TMR sensor. The applied bias field was chosen for

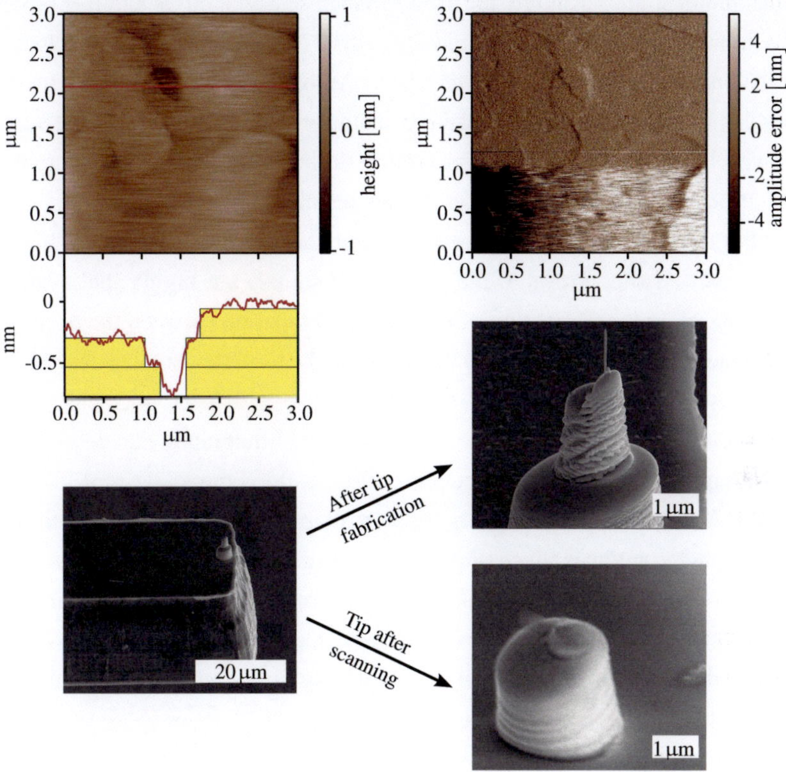

Figure 4.5.: With manually grown tips by focused ion beam deposition, the lateral resolution of the cantilevers is significantly improved. Amplitude modulation mode imaging with the feedback on the TMR sensor reveals atomic step-edges on gold(111) terraces. In the lower section of the scan image however, it appears that the tip geometry or cantilever properties have changed significantly and disable a stable imaging of the sample. As shown in the SEM micrographs (Image courtesy: Dirk Meyners and Ali Tavissolizadeh, CAU Kiel), the tips are grown on the apex of the cantilever. The sharp tips ($r_{Tip} = 30\,\mathrm{nm}$) however can easily break during scanning as revealed by SEM.

maximum strain sensitivity for the unstrained sensor at 60 Oe and $\alpha = 115°$. With those parameters, the atomic step edges of 2.54 Å height are resolved. The image was scanned downwards from the top. As the apex of the tip used in this scan was a fine rod of 30 nm in diameter and 500 nm in length, it was a very unstable tip. In the lower section of the image, the tip geometry is significantly changed and makes the last scanlines blurry.

For dynamic mode experiments, not only amplitude signals are of high interest. Also contrasts in the phase shift signals are highly interesting as they provide information of the energy dissipation between the tip and the underlying material[152] and therefore can be used to visualize chemical contrasts[153] (see Chap. 2.1.2). To pattern samples with low topographic contrast but high chemical contrast, polymer blend lithography like described in Chap. 2.2.1 was used. Especially self-assembled monolayers of FDTS ($1H,1H,2H,2H$ - perfluorodecyltrichlorosilane) are well known for their hydrophobic surfaces. In contrast, SiO_x is very hydrophilic compared to the FDTS-SAM. The SAM-formation and patterning by polymer blend lithography of FDTS on SiO_x wafers is also well documented in literature.[51] The FDTS hereby form a monolayer of 1.3 nm height. This surface morphology has a strong influence on the thickness of the water film[22] on those surfaces. If exposed to ambient conditions with a relative humidity of around 40 %, topographic contrast on those sample systems disappear in amplitude modulation imaging due to the various thickness of the water films on FDTS and SiO_x and, therefore, a huge variation in capillary forces. However, as shown in Fig. 4.6 a) the difference of the energy dissipation between the two materials is still measurable. On this sample, the holes in the FDTS-SAM are visible as bright spots in the phase signal. Additionally, the quality factor of the cantilever used for this measurement, was extremely high with values around 1200. As discussed in Chap. 2.1.2, for samples with high contrast for dissipative tip-sample interactions and high quality factors, frequency modulation imaging can be an advantage.

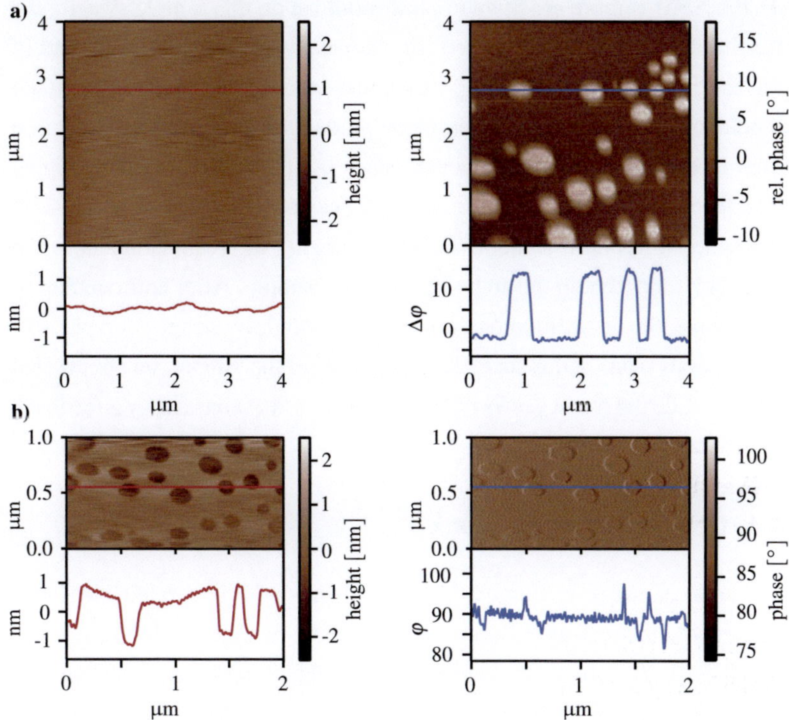

Figure 4.6.: Dynamic mode imaging of FDTS-SAM with TMR sensor with the feedback on amplitude and phase. **a)** Amplitude modulation mode imaging of FDTS-SAM in SiO_x with TMR sensor. The SAM has been patterned by polymer blend lithography. As the SiO_x is very hydrophilic and the FDTS in contrast is very hydrophobic, the topography of such samples are normally invisible to amplitude modulation AFM, especially if the sample is exposed to environments with a humidity above 40 %. However, the phase contrast still reveals the different materials of the sample due to different energy dissipation between tip and sample. **b)** As the dissipation is the more sensitive signal, frequency modulation AFM can reveal the topography of the sample. As the cantilevers resonance frequency is feed back to the driving signal by an additional loop, the phase contrast vanishes and is constant at $90°$, while the topography with the holes in the SAM is revealed.

As the AFM is more sensitive to phase contrast on this sample system, the feedback loop can be readjusted for a constant frequency shift instead of the amplitude feedback. In Fig. 4.6 b), the same sample used in amplitude modulation was scanned in frequency modulation. As the self-excitation loop feeds back the cantilevers resonance frequency to the driving signal by a feedback loop on the phase signal to keep the phase at $90°$ phase shift; the contrast in phase signal disappears. During the readjustments in the feedback loop, the tip has to be lifted of the sample. After approaching to the sample, the tip position on the sample was slightly changed compared to the previous scan. In this scan, the topography of the sample can be revealed clearly while the phase shows no contrast as it is kept constant by a feedback loop during scanning.

4.4. Conclusion and Outlook

In summary, micro-machined AFM cantilevers with integrated TMR sensors enable the self-detection of the cantilever bending with a sensitivity of $3.9 \times 10^{-6}/\text{Å}$ which is superior to piezoelectric and piezoresistive sensors. The successful recording of force vs. distance curves and topography images demonstrates that the classical optical beam deflection method can be replaced by this type of cantilevers. The sensitivity of the TMR sensor can be tuned by a magnetic bias field and adjusted for optimal sensitivity for the desired operational mode. In contact mode AFM, this behavior can be utilized to adjust the working point of the sensor in a way, that it shows the highest sensitivity for a strained state when the tip is approached. Additionally, strain sensitive TMR cantilever are also suitable for dynamic operation in amplitude modulation as well as for phase locked experiments. By growing tips by electron beam deposition to the cantilevers, a high spatial resolution is achieved. Additionally, atomic step edges of gold(111) and self assembled monolayers can be imaged with such cantilevers. However, ongoing studies have to extend the application of TMR-based self-sensing cantilevers to other environments (liquids and vacuum) and eliminate the current need of a magnetic bias field during operation.

A shortened version of the first part of this chapter has been published as the article "Self-Sensing Atomic Force Microscopy Cantilevers Based on Tunnel Magnetoresistance Sensors", A. Tavasollizadeh, T. Meier, K. Rott, G. Reiss, E. Quandt, H. Hölscher and D. Meyners: Applied Physics Letters **102**, *153104 (2013).*

5. Thermal Conductance of Chain-Like Molecules

For thermal analysis on the nanoscale, scanning thermal microscopy (SThM) offers the advantages of the high spatial resolution of a scanning probe method. In modern microprocessors structural details are too small to be resolved with optical microscopy techniques. Intel microprocessors, for example, fabricated on the 22 nm node went on sale in April 2012 and will be replaced by processors fabricated on the 14 nm node in 2014. Interestingly, the limiting factor in performance of modern processors is the dissipation of heat and, therefore, energy efficiency. Thermal probing of such devices helps to improve performance and reliability. Photothermal spectroscopy and microscopy, however, have to break the diffraction-limited barrier in optical imaging to enable insights into thermal properties of today's microprocessors. With far field techniques like time-domain thermoreflectance,[154] the diffraction limit is the barrier for spatial resolution. In nonmetallic materials, heat is transported by phonons with a wide variation in frequency and an even larger variation in their mean-free-path of $1 - 100$ nm at room temperature. In modern micro- and nanostructures, the phonon mean-free-path is, therefore, in the same scale or even larger than structural features.[155] This rises interesting questions on transport mechanisms and more fundamental on the definition of temperature at these scales.

In the field of heat transport, one of the most controversial topics is on finding analytic descriptions of heat transport in periodic nanoscale structures. Even by simplifying the geometry to one dimensional linear spring mass systems, transfer mechanisms, which were first investigated by Schrödinger in 1914[14] and Peierls in 1929[15] during the last century, are still frequently studied. In the discussion on transfer mechanisms, one-dimensional chains of springs and masses are peculiar systems. Although they are many-particle systems, they do not behave like thermodynamic ensembles since they do not equilibrate. This leads to unexpected behavior also in terms of thermal transport like the Fermi-Pasta-Ulam paradox.[156] Specifically, 1D chains do not follow Fourier's law,[157, 158]

$$\dot{Q} = \bar{k}\, \nabla T\,, \tag{5.1}$$

equating the heat flux density $\dot{Q}$ to the constant thermal conductivity $\bar{k}$ times the temperature gradient ∇T. In the classical 3-dimensional case, the thermal conductivity thereby is defined as the amount of heat flowing through a plate of a particular area and thickness at a given temperature gradient between the opposing areas (units W/m· K). However, by only discussing Fourier's law in a 1-dimensional case, the heat flux cannot be normalized on a particular area and the unit of the thermal conductivity is reduced to W· m/K. Especially, for linear (1D) chains, it is found that the thermal chain conductance diverges with chain length $L = a\cdot N$, with N as the number of elements in the chain and a as the distance between two elements, as $k \propto N^{\alpha}$, with $\alpha > 0$, instead of being a material-intrinsic property as required for Fourier's law ($k = $ const. or $\alpha = 0$). After extensive investigation,[15, 157, 158] it was concluded from simulations that $\alpha = 1/3$ is a typical value for long chains ($N > 100$). The thermal conductance G_m (units W/K) of a single chain,

$$\dot{Q} \quad \propto \quad N^{\alpha}\Delta T\,, \tag{5.2}$$

$$\Rightarrow G_m \quad = \quad \frac{\dot{Q}}{\Delta T} = \frac{k}{L} \propto \frac{N^{\alpha}\Delta T}{a\cdot N\Delta T} \propto N^{\alpha-1}\,, \tag{5.3}$$

scales with the 1D-thermal conductivity k, where L is the length of the chain increasing with N. Therefore, $G_m \propto N^{\alpha-1}$. Extending this analysis to shorter chains is of high technological relevance.[159] For example, self-assembled monolayers (SAMs) of alkane chains on surfaces, in particular on silicon chips or gold electrodes anchored using trichlorosilane[160] and thiol end groups,[161] and molecular systems are excellent candidates for surface modifications to enable electronic and sensor applications as well as energy conversion, where thermal transport is of high relevance.

Simulations of short chains ($N < 100$) indicate that the coupling strength between the chains and the thermal reservoirs governs[157, 162, 163, 164] α-values ranging from 0 to 1. Indeed, a strong dependence of the conductance on the coupling of molecular chains (SAMs) to one[165, 166] or two thermal reservoirs[167, 168, 169] has been demonstrated experimentally. However, the length dependence of 1D (or molecular) conductance for short chains is yet to be fully understood. Molecular dynamics simulations of SAMs[164] predict significant phonon interference effects with conductance variations of about a factor of 3 as a function of N, for $N = 3, 8, 16$ and 32. In a more detailed quantum-mechanical simulation, Segal et $al.$[162] also predict conductance variations in the same order of magnitude, but with a single maximum at $N = 4$ and approaching a constant conductance for $N > 20$ for a weak coupling of the molecules to the thermal reservoirs and $\alpha \approx 0$ for a strong coupling. In contrast, Duda et $al.$[170] argue that the thermal conductance is fully ballistic and, therefore, is only an interface effect to the contacts and independent of length ($\alpha = 0$) for $N \geq 5$ because of a length-independent density of phonon states.

Experimental data of SAMs sandwiched between gold and gallium arsenide surfaces[167] or between gold and water[165] is too scarce to allow the extraction of a length dependency. The seminal time-resolved flash heating experiments of SAMs through a gold substrate[166] ($N = 6$ to 24) were fitted using $k^{-1} \propto (N-3)$ for $N > 5$, implying $k^{-1} \approx 0$ for $N \leq 3$ (no thermal resistance at all). However, the opposite trend was obtained when measuring and simulating intramolecular equilibration times through alkane linkers;

namely, a Fourier-like trend ($\alpha = 0$) for $N \leq 4$ and constant conductance ($\alpha = 1$) for $N = 5$ to 8.[171]

With this vast variation of models and experiments, measuring the thermal conductance along short alkane chains seems like a tempting experiment for testing the limits of the thermoresistive sensing. Additionally, short alkane chains are a well understood model system to study electrical transport properties with scanning probe methods[172] and are in various configurations commercially available.

5.1. Scanning Thermal Microscopy

On nanoscopic length scales, material properties and transfer mechanisms can differ from bulk materials. With its high spatial resolution, scanning probe microscopy has become a powerful tool to investigate mechanical, chemical, electronic, magnetic and in this case thermal properties on length scales not accessible by other methods. In electric transport measurements, scanning probe based measurements have shown to be a valuable tool for nanoscale electrical characterization,[172] as they can also resolve sample defects. In contrast to electrical transfer, which is limited to materials with electric carriers, the thermal transfer of energy occurs in all materials. Additionally, the second law of thermodynamics requires the dissipation of energy in the form of heat for all irreversible processes that involve energy interactions with their surroundings. This is the basic principle of many physical phenomena like magnetic hysteresis, phase transformations and electric heating. Because energy in the form of heat can be transported in all states of matter and by various transport mechanisms, controlling heat flux especially in nanoscale devices like microchips still remains hard and is the limiting factor in the evolution of performance.[173] Controlling the energy flux via heat is, therefore, crucial for the measurement of temperature and energy transfer. In nanoscale devices, this requires both high spatial and temporal resolution. Modern time-resolved flash heating experiments enable a high temporal resolution thanks to femto-second laser pulses, but with a poor spatial resolution[166] due to the diffraction limit. Scanning thermal microscopy (SThM)[11] is adapted from atomic force microscopy and combines therefore its high spatial resolution[12] with a intermediate temporal resolution.[9]

Like all scanning probe microscopes, scanning thermal microscopy relies on a sharp tip which is used to probe a samples surface mechanically. Compared to atomic force microscopy, however, scanning thermal microscopy is highly specialized on probing specific tip-sample interactions, namely the heat flux between the sharp tip and the sample. The scanning thermal microscope used for the experiments in this chapter, therefore, was designed for high precision measurements of the heat flux by Fabian Menges *et al.*[13] using a thermoresistive probe integrated into an atomic force microscopy cantilever at the IBM-Zurich Noise-Free-Labs.[174] With the sample mounted on a $10 \times 10 \times 10\,\mu m^3$ piezo scanner, this allows a high spatial resolution in thermal means even without a force feedback mechanism to control tip-sample forces because of a missing optical read-out of the cantilever's deflection in the vacuum operated instrument.

5.2. Experimental Details

5.2.1. One-Dimensional Character of SAMs

To interpret thermal transport in the measurement's geometry of this experiment (and similar ones described in the literature) there are two questions to be answered. First, can inter-molecular transport be neglected compared to transport along the molecular chains? Secondly, is the measured conductance over a small ensemble of molecules probed simultaneously in the experiment equivalent to the sum of identical molecular contributions?

To answer the first question, the geometry of the experimental arrangement and the conductivity ratio must be considered. The molecules in the SAM are relatively upright with an angle of $28°$ with respect to the surface normal. This angle increases only moderately by about $10°$ in the pressure range applied in this experiment.[175] Thereby, through all experimental data, it is considered that the molecules in the contact zone touch both tip and gold surface. Next, the thermal conductance ratio is considered independent from the data obtained in this study. Regardless of whether molecular films are crystalline (as in SAMs) or amorphous (as in polymers), the thermal transport is found to be dominating along the molecular chains (intra-molecular) with an at least one order of magnitude smaller conduction between molecules (inter-molecular),[170, 176, 177] justifying their treatment as quasi-1D conductors when the direction of heat flow is along the chains. The thermal conductance along crystalline polyethylene (PE) along the molecular direction showed[177] a thermal conductivity on the order of 100 W/Km, more than two orders of magnitude larger than in amorphous PE in which intermolecular transport must be an integral part of transport. This is consistent with earlier work measuring anisotropy of thermal conductance in stretched polymer films[178] resulting in an estimate of a conductance ratio of inter- versus intramolecular conductivity of at least ten. A high conductivity ratio is plausible because of the difference in bond strength of van der Waals versus chemical bonds.

Secondly, to be able to treat the measured conductance per area as composed of individual and independent conductance channels per molecule as predicted from single-molecule simulations, the lateral coherence length l_c of phonons entering the solid-SAM interface from the bulk must be smaller or comparable to the lateral spacing between the individual molecules to ensure transport along a single molecular chain. With this assumption, the measured thermal conductance per area G_A can be expressed as a sum of the thermal conductance per molecule over the number of molecules within the contact area $G_A = \sum_{n=1}^{N} G_{m,n} = N \cdot G_m$ with the individual thermal conductance $G_{m,n}$ of the n-th molecule in contact and N molecules in contact in total. This indeed appears to be the case: using the velocity of sound in gold of $c = 3240$ m/s, and the temperature $T = 300$ K, the coherence length[179] in gold is estimated to $l_c = hc/(k_B T) = 0.52\,$nm. For silicon, the speed of sound is $c \approx 6084\,$m/s and the temperature about $T \approx 520\,$K and therefore $l_c = 0.56\,$nm. These lengths translate into an area of $A = \pi l_c^2/4 = 2.4 \times 10^{-19}\,$m^2 and $2.1 \times 10^{-19}\,$m^2 for coherent phonons in silicon and gold, respectively. In comparison, the area per molecule in the SAM is $A_{\mathrm{SAM}} = 2.14 \times 10^{-19}\,$m^2. The magnitudes are similar, and the expected error calculating G_A directly into G_m is, therefore, small compared to the systematic errors, even if more accurate estimations of l_c are used.[180] G_A is thereby the thermal conductance measured per area (within the mechanical contact between tip and sample) $G_m = A_{\mathrm{SAM}} \cdot G_A$.

5.2.2. Contribution From Tip and Substrate to the Thermal Resistance

As the tip-sample heat path consists out of 3 thermal resistors in series, $R_{Tip}, R_{SAM}, R_{substrate}$ (see Fig. 5.2 for details), the contributions of tip and substrate to the measured thermal conductance have to be taken in account. The thermal conductance of tip and substrate, respectively, are estimated using the following assumptions. The contact radius calculated using the contact model described in Chap. 2.1.1, is on the order of up to $r = 5\,$nm. The conductance due to thermal spreading in gold (assuming diffusive transport of electrons) with $k = 318\,$WK^{-1}m^{-1} at room temperature can be calculated to be $G_{th,Au} = 4rk = 6.25 \times 10^{-6}\,$WK^{-1}, which is about three orders of magnitude larger than the experimental value of G_{th}. One may argue that due to the weak electron-phonon coupling, only the phonon transport matters at these length scales. Therefore, the phonon contribution to the conductivity on gold is estimated using the electrical conductivity $\sigma \approx 4 \times 10^7/(\Omega\,m)$ and the Wiedemann-Frantz law: $k_{electronic} = LT\sigma = 292\,$WK^{-1}m^{-1} using $L = 2.44 \times 10^{-8}\,W\Omega$K^{-2}. The conductivity $k_{phonon} \approx k - k_{electronic} \approx 26\,$WK^{-1}m^{-1} therefore corresponds to a phononic spreading conductance of $\approx 5 \times 10^{-7}\,$WK^{-1}. From the Boltzmann transport equation one can estimate a phonon mean free path of $\Lambda \approx 10\,\text{nm}$. Therefore, one can also use the ballistic approximation $G_{Au,phonon} = \frac{3\pi k a^2}{4\Lambda} \approx 1.4 \times 10^{-6}\,\text{WK}^{-1}$, which is still about two orders of magnitude larger than the measured value of G_{th}.

A similar argument applies to silicon. Here, the tip diameter sets the expected phonon thermal conductivity in the quasi-ballistic regime[181] on the order of $35\,\text{WK}^{-1}\text{m}^{-1}$, which is on the same order as the gold phonon thermal conductivity. Accordingly, very similar thermal resistances are estimated.

In conclusion, the expected influences from neglecting the thermal resistance of silicon tip and gold substrate are smaller than the scatter in the experimental data.

5.2.3. Experimental Setup

The experiments were carried out in the noise-free-labs of IBM-Zurich with the scanning thermal microscope designed by Fabian Menges *et al.*[9, 12, 13, 134] operated in vacuum conditions with heatable cantilevers. For the thermal transport measurement, the heater/sensor in the cantilever represents one thermal reservoir at the temperature T_{heater}, while the sample is thermally connected to the vacuum chamber at room temperature. Operated at ambient conditions, the conductance of the air gap between tip and sample is strongly distance dependent and can be used as a topography signal for the sample when scanning the tip close to the surface.[182, 183] This allows the recording of the individual tilt and position of the sample before evacuating the chamber. Operation at high vacuum ($< 10^{-6}\,\text{mbar}$) is required because the thermal conductance of air due to diffusion dominates the heater/sensor signal when operated at ambient conditions. At vacuum conditions, no distance dependence of the thermal conductance of the gap can be observed because of the absence of air. Therefore, the vacuum assures that the heat from the heater/sensor is either transported through the cantilever or the tip and sample to the thermal reservoir at room temperature (20 °C). By monitoring the electrical power used for keeping the heater/sensor at constant temperature, the heat flux through both thermal transport channels (the cantilever and the tip-sample contact) can be observed. After the tip has been brought into contact, it can be scanned in constant height mode using the previously recorded sample tilt in a feed forward loop. However, to achieve a sufficient force control, the samples used in the experiments have to be as flat as possible since the missing force feedback makes it impossible to control varying forces due to the sample's topography.

5.2.4. Scanning Thermal Microscopy on SiO_x Substrates

To ensure the tip's integrity during a measurement, a single scan technique is required for a quantitative analysis of the measured data. As the heat flux is predicted to be very sensitive to variations in the coupling strength to the thermal reservoirs[165, 166] and scanning tips tend to wear out,[184, 185] the control of the contact area is crucial for a quantitative analysis.[91] The easiest way to prove the tip's integrity is, therefore, to scan a sample with different materials and look for material contrasts (see Fig. 5.1 a)). This requires a unique patterning of the sample to distinguish between different materials just by the shape of the patterning. To create a characteristic pattern on a flat surface, polymer blend lithography on silicon wafers with native oxide films fulfills the requirements for the fabrication of flat samples with characteristic patterns.[51] Like described in Chap. 2.2.1, this technique holds the advantage to fabricate polymeric masks on silicon wafers with well defined geometry by a single spin-coating step without the need of further lithography. After depositing the trichlorosilane-terminated alkanes forming a self-assembled monolayer, the remaining mask is selectively removed by a snow-jet treatment and a patterned SAM remains on the substrate.

Additionally, the trichlorosilane SAMs also allow to functionalize the SiO_x tip of the scanning thermal microscope enabling a symmetric sandwich

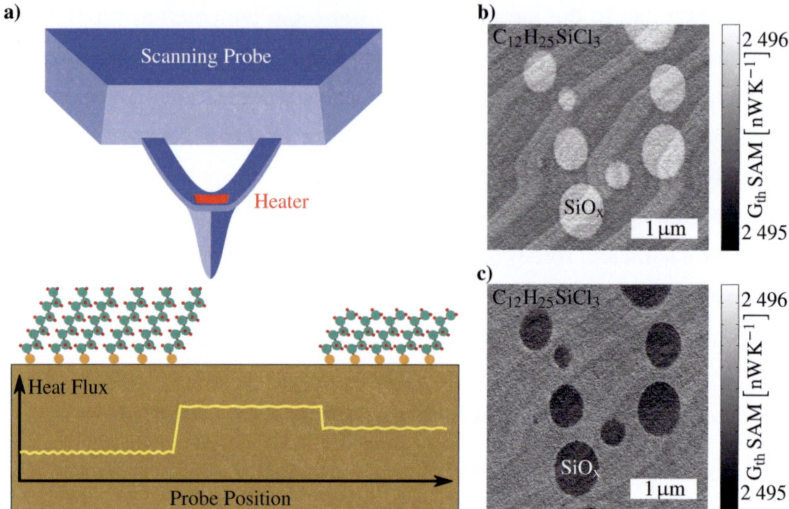

Figure 5.1.: a) Schematic of scanning thermal microscopy on self-assembled monolayers. While the tip is scanned along the sample, the heater's temperature is kept constant while the heat flux between the tip and the sample is monitored. The blank substrate is used as a reference while the length of the alkane chains between substrate and tip is varied. **b)** An example of a thermal image of dodecane-trichlorosilane on SiO$_x$. The two phase dodecane-SAM template is showing a clear thermal contrast between SiO$_x$ and the SAM. **c)** Contrast inversion after recording 2 more pictures on the same position of the sample. As the sliding tip picked up some material from the sample, the tip configuration changed irreversibly because of the SAM molecules bonding to the SiO$_x$-tip of the microscope.

of SiO$_x$-SAM-SAM-SiO$_x$. In a first step, the sensitivity of the scanning thermal microscope to the proposed experiments was shown by scanning two phase alkane-SAM/SiO$_x$ templates. Preliminary tests on templates with alkane chain lengths of 4 to 18 methylene units reveal material contrast between the SiO$_x$ and the SAM on all samples.

One example of such a thermal conductance measurement on a dodecane-SAM template is given in Fig. 5.1 b). However, the quantification of the obtained data is still not possible, as the tip integrity is indeterminate. Like previously discussed, a sliding tip wears out, scratches the surface or pics up material from the surface. Those effects have been observed on all SAM/SiO$_x$ templates and result in a change of the material contrast and in extreme cases to a complete inversion of the observed contrast. In Fig. 5.1 c), the material contrast observed in Fig. 5.1 b), has been inverted after two more scans on the same position. This contrast inversion is most likely related to a chemically modified tip, as the contrast remains when changing to a different position on the same sample. The most likely scenario explaining this inversion is the breaking of some bonds anchoring the SAM's molecules on the SiO$_x$ and releasing the molecules to the tip. As the molecules adsorb on the tip, they also can form a SAM-like layer on it with a unknown molecule density which influences the heat flux between tip and sample. The tip contamination is also permanent, which indicates a chemical bonding to the tip. By heating the tip up to 1000 °C, all unbound organic contaminants simply burn away which did not restore the tips' initial configuration after a contamination on the silane-SAMs.

In summary, the simple fabrication of the templates offered a fast and cheep approach for the fabrication of model systems to study heat transfer along molecular chains. The experiments on the two phase alkane-SAM/SiO$_x$ templates show encouraging results in the terms of sufficient sensitivity to material contrasts. However, as the used chemistry to immobilize the molecules on the substrate was not selective only to the substrate, but also to the tip of the microscope, it could not be guarantied to only contact a monolayer of molecules. These varying preconditions during the experiments, causing non-reproducibility due to tip contamination, only allow qualitative conclusions.

5.2.5. Thermal Force Mapping on Au(111) Substrates

As seen during the measurements on the silicon wafer substrates, controlling the contact geometry and coupling between the tip and the SAM is a major challenge when performing thermal conductance measurements. In addition, tip wear and contamination must be avoided. To avoid contamination and tip wear, two parameters can be varied. First, the chemistry which was used to bond the molecules on the substrate was changed and secondly, the operational mode of the scanning thermal microscope was modified. To change the bonding chemistry, a different set of molecules with a thiol-endgroup (instead of trichlorosilane-endgroups) on one end of the alkane chain, which can specifically bind to mintage metals, was chosen because they will not chemically bind to the silicon tip. This also requires an exchange of the substrate. As a substrate material for the thiol molecules, gold is the natural choice for the proposed experiments. Although structuring patterns in thiol SAMs on gold has been demonstrated before and can be done by microcontact printing (μCP)[186] or atomic force microscopy,[187] the introduction of thermal force mapping (see Fig. 5.3) makes any patterning obsolete. Additionally, if grown on (atomically) flat mica substrates, the gold formed crystals with the characteristic Au(111) terraces facing towards the tip of the scanning thermal microscope. In contrast to the amorphous SiO_x substrates, the crystalline structure of the Au(111) is very regular, enabling a regular crystalline formation of the SAM on the terraces with a well defined density of molecules.

The use of different molecules on the Au(111) substrates successfully avoids chemical bonds between the tip and molecules and allows the introduction of thermal force mapping. This technique requires repetitive contacting of the SAM with the tip and lifting of the tip from the SAM without chemical bonds between tip and the molecules of the SAM. Here, the alkane molecules are sandwiched between a gold-thiol bond (strong coupling) and the silicon SThM tip (weak coupling), see Fig. 5.2 a). This allows the use of repeated contacting/detaching cycles of the tip and the surface rather than scanning images to avoid wear, achieve excellent reproducibility and obtain sufficient sensitivity.

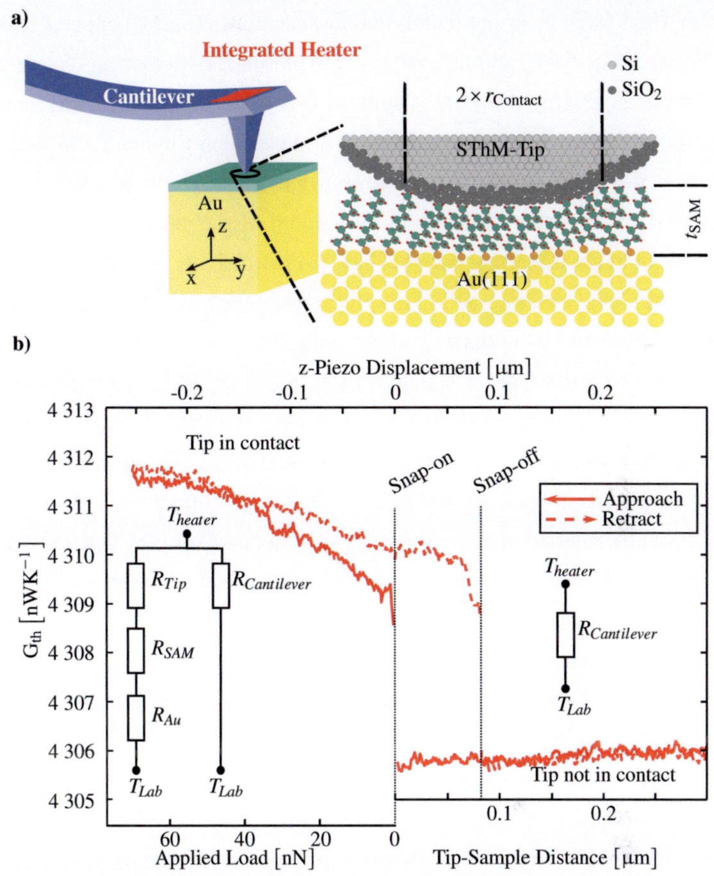

Figure 5.2.: a) Schematic of the contact geometry. The contact radius, r_{Contact}, is a function of the applied load and the thickness of the SAM, t_{SAM}. **b)** Example of thermal conductance versus displacement measurement on octadecanethiol. During contacting the sample with the tip, the heat flow from the heated tip through cantilever and sample is recorded. By analyzing the noncontact and the contact regime of the thermal conductance vs. displacement curves and the geometry of the tip-sample contact, the thermal conductance of the SAM can be extracted in a space-resolved manner. The applied z-piezo displacement (top axis) translates into a variation of the tip-sample distance or the load force when the tip is not in contact or in contact with the surface (bottom axes), respectively.

The thermal force mapping technique now combines thermal contact curves as shown in Fig. 5.2 b) and recording of the thermal conductance, G_{th}, versus the displacement of the piezo-element carrying the cantilevered sensor with the tip. Because of the weak coupling between tip and SAM and the absence of shear forces, the tip can be lifted without contamination from the SAM. With a stepwise lateral movement of the sample under the tip between individual thermal conductance-versus-displacement curves, a high spatial resolution is obtained. The procedure is related to so-called force-mapping by force-vs-distance curves using scanning force microscopy.[20] The cantilever's thermal resistance is extracted from the measured response of the heater sensor without tip-surface contact. The additionally measured conductance with the tip in contact is assigned to the tip-surface heat path. With a stepwise lateral movement of the sample under the tip between individual thermal conductance-versus-displacement curves and analyzing the individual thermal contact curves in a space-resolved manner, maps of mechanical and thermal properties can be generated.

Compared with conventional surface scans, approach and retraction of the tip to and from the sample greatly simplifies the contact mechanics because of the absence of friction forces (the contact areas of sliding tips are typically distorted with respect to mechanical contact models). Furthermore, avoiding shear forces occurring during sliding motion helps to minimize wear and contamination and, thereby, preserves the integrity of tip and surface for subsequent measurements. This also enables the application of well understood contact models to the SAM-tip interface to describe the contact mechanics. By analyzing the individual thermal contact curves carefully, additional information of the sample properties can be extracted. For instance, a topographic image of the sample can be generated even without a force feedback loop. Analyzing the load dependence of the thermal conductance facilitate an understanding of the limits of the experiment like the integrity of the crystalline structure of the SAM or identifying defects in the substrate. As these limitations of the experiment can be directly

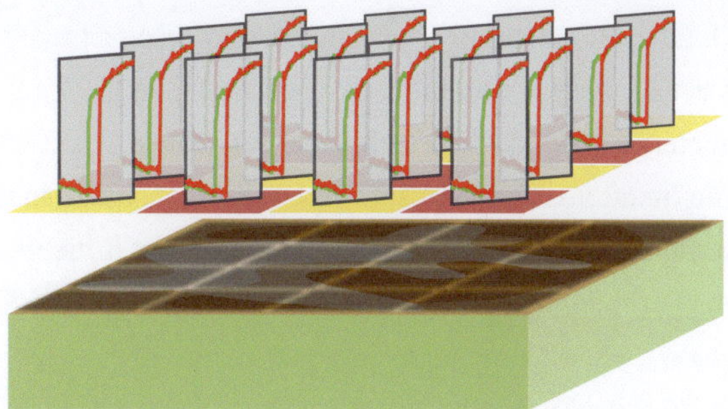

Figure 5.3.: Schematic of the thermal force mapping technique: This technique measures highly accurate thermal conductance versus displacement curves in a 2-dimensional array on the sample. The applied forces can be controlled and the offline analysis allows the quantitative property calculation of topography, tip-sample adhesion contact area and thermal conductance.

identified in the dataset, the interpretation of such a set of data is also greatly simplified and allows the quantification of the thermal conductance for alkane-chains as a function of their chain length as demonstrated in Fig. 5.6. To calculate the contact area, the adhesion force has to be measured, During the retract of the tip from the sample, it can be observed directly by the snap-off of the cantilever. As the force from the spring constant of the cantilever is strong enough to pull the tip off the sample, the distance of this snapping off with respect to the contact point during approach (snap-on) can be used for the calculation of the adhesion force. The snap-off position, however, has a higher experimental scatter than the snap-on position. One of the reasons for the scatter of the snap-off position measurement is the dependence of decohesion on the state of the molecules of the SAM. When the tip-surface force reaches a certain threshold during the load cycle, the crystalline structure of the film is broken by the high pressure to yield the tip.[188] This changes the contact geometry drastically and, therefore, the tip-sample interaction, which can easily be identified in the thermal contacting curve. Only after complete decohesion of the tip, the molecules rearrange into their original crystalline structure. As a result, the indentation curves (during the load cycle) are more reproducible and easier to analyze than the retraction curves, which is why the analysis is based on the load cycles only.

All parameters of Eq. (2.15) are either experimentally accessible or can be taken from literature. The largest uncertainty is contained in the value of the Young's modulus of the SAM, E_{SAM} which is very difficult to measure. Comparing experimental and theoretical analysis of SAMs,[175] $E_{SAM} = 40 \pm 20\,\text{GPa}$, and $\nu_{SAM} = 0.42$ appears to be a good estimate.

For high load forces, a threshold pressure is reached and the tip eventually starts to punch through the SAM and touches the gold substrate directly. To limit the pressure, a relatively blunt tip with a radius of 28 nm was used to apply several tens of nN before the punch through. The punch-through can also be identified directly in the recorded data by a kink in the measured thermal conductance at a characteristic force threshold of several GPa.[175, 188] Therefore, there exists an upper limit for the force applied that allows assuming an ordered film in the analysis of the thermal contacting curves. This maximum force can easily be determined in the thermal contact curves as they tend to change the shape of the curve dramatically. By breaking the crystalline structure of the SAM, the tip penetrates into the SAM increasing the contact area,[188] and this scenario can be observed by a step-like increase of the load-dependence of thermal conductance (see Fig. 5.4).

As all materials conduct heat there is a permanent flux of energy from the heated region in the cantilever via the cantilever to the lab environment. As shown in Fig. 5.2 b) the heat flux is only along the cantilever as long as the tip is not in contact with the sample. The temperature of the heater can be calibrated with the procedure described in Chap. 3.3 and the thermal conductance of the cantilever, while not in contact with the sample, can be extracted using Eq. (3.17). The thermal conductance of the cantilever is, therefore, used as a reference to quantify the tip-sample heat path as the tip is brought into contact with the sample.

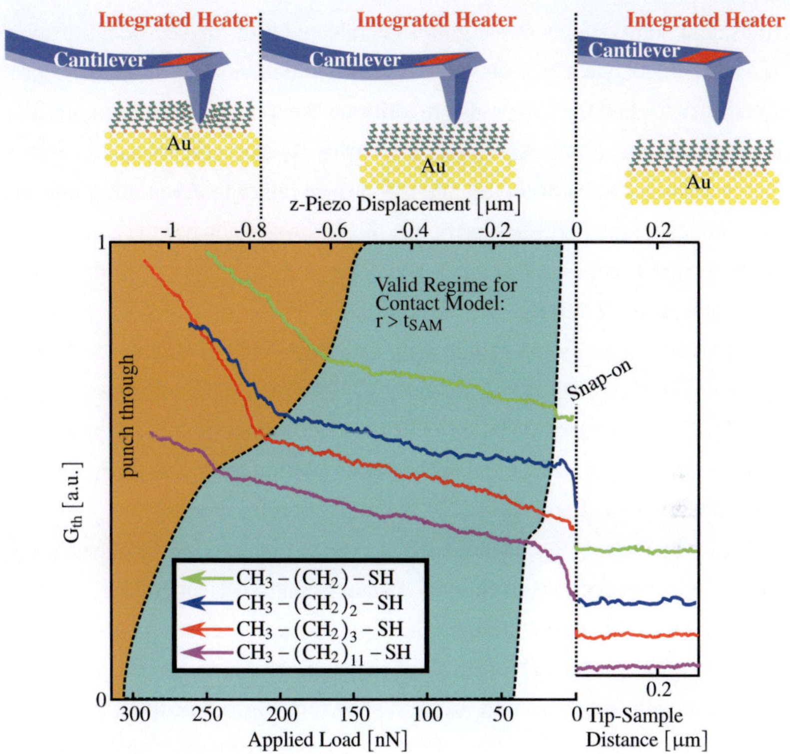

Figure 5.4.: Thermal contact curves with large loading forces. As the load between tip and sample reaches a certain threshold, the tip breaks the crystalline structure and elastically[189] indents the film. The contact geometry changes drastically for forces higher than the threshold which can easily be identified by the abrupt increase in the measured thermal conductance. This defines the upper boundary in terms of applied load for each individual chain length.

The goal of thermal force mapping experiments is to shed light on the dominant transport mechanisms along short alkane molecules as a function of the molecule length. For such studies, SThM offers the advantage of a local imaging method for simultaneously mapping the samples' topography and heat flux. This feature allows the imaging of the samples' topography[11] and mechanical properties to identify local defects, such as grain boundaries, and point defects of the gold substrate and film. In electrical transport measurements, such defects have been identified as a cause of nonreproducibility,[172] and they may also deteriorate the quality and reproducibility of thermal transport measurements on SAMs. By using the repeated contacting/detaching cycles of the tip and the surface rather than scanning images, as obtained by thermal force mapping, this technique can use the cantilever's thermal resistance as a reference and achieve an excellent reproducibility with sufficient sensitivity.

The samples analyzed were highly ordered SAMs of alkane thiols (HS-$(CH_2)_{N-1}$-CH_3), with $N = 2$, 3, 4, 8, 12, 14, 16 and 18 methylene units, on Au(111) substrates on Mica. The molecular monolayer was grown by immersing the substrates into 1 mM ethanol solution following the procedure described by Delamarche $et\ al.$[190] to obtain uniform monolayers.

However, thermal force mapping expands the time needed for one map significantly compared to conventional scanning and the density of the 2D-array is limited by the instrumentation time. For the measurements presented here, contact curves on 60×60 locations of the surface of each sample separated by 16 nm were recorded and maps as shown in Fig. 5.5 of the topography (Snap-on), the adhesion force (difference between Snap-on and Snap-off multiplied by the cantilever's spring constant), the load force dependence (slope of the contacting part) and the thermal conductance of the sample with high spatial resolution were calculated. This allows to identify and discard data taken in regions with step edges and defects of the sample surface.[20] To obtain the thermal conductance per molecule, measurements with sensor/tip temperatures of 200 °C and 300 °C and peak forces keeping the integrity of the tip and SAM intact were recorded on multiple positions on each sample. After discarding the results for areas with defects on the samples, the remaining thermal contact curves were analyzed and combined into the thermal conductance maps for each individual chain length. Compared to flash heating experiments which reach approximately 800° C initially, the temperatures used here are small.[166] Wang $et\ al.$ also found a strong temper-

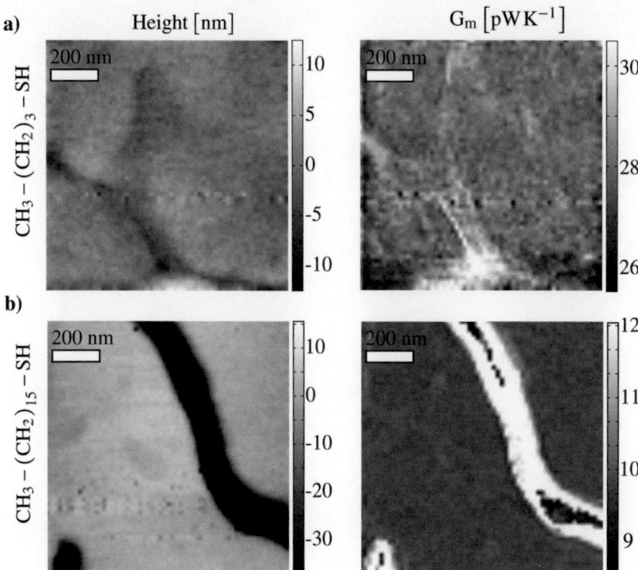

Figure 5.5.: Example of topography and thermal conductance maps obtained for two different SAMs. The high spatial resolution of scanning-probe-based thermal mapping allows the identification of topographic and defect artifacts in the thermal conductance measured that cause nonreproducibility. In **a)** the typical (111) terraces of gold can be resolved and no cross talk to the thermal map can be observed. In **b)**, however, a grain boundary between two gold islands was recorded and can be excluded from further analysis as it is a substrate defect on the sample.

ature dependence for heat transfer along alkane chains bonded on gold by thiol groups for temperatures up to 200° C.[167] Above 200° C however, they found the influence of the temperature on the conductance to be reduced and with rising temperatures no influence has been observed. Therefore, also the data for both temperatures measured here show no difference in the observed trend.

The thermal conductance measured (see Fig. 5.6 a)) varies with the piezo position as the resulting contact force enhances the tip load on the surface and subsequently increases the contact area. To obtain the thermal conductance

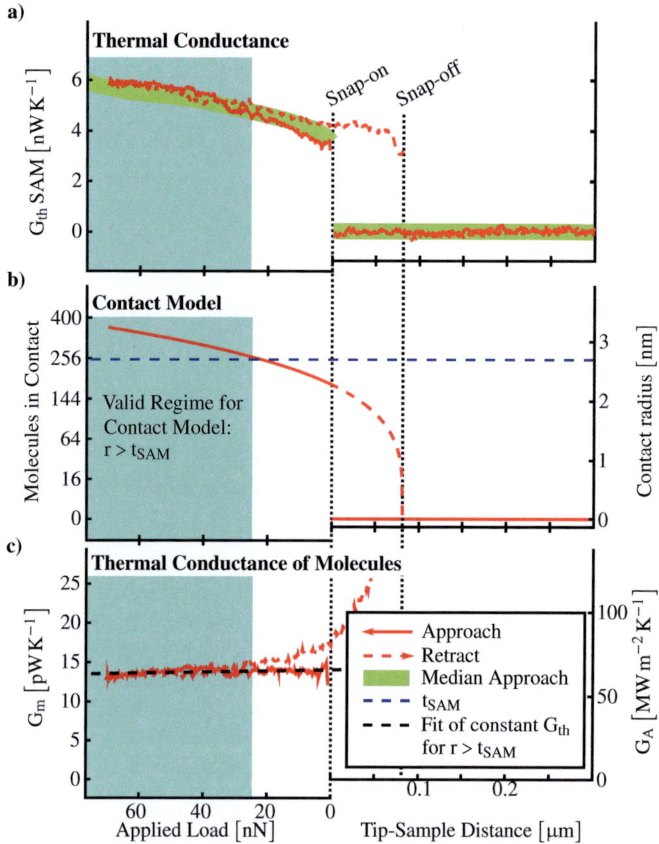

Figure 5.6.: Example of analysis of the molecular thermal conductance using a contacting curve of an octadecanethiol SAM. **a)** Results of translating the contacting curve of Fig 5.2 b) into conductance versus applied load. Also shown is the average over 2500 such curves. The contact model applicable to this sample is plotted in **b)**; the shaded region denotes the region of validity of the model. Combining the data of a) and b), the thermal conductance per unit area, G_A, or per molecule, G_m, can be extracted, as shown in **c)**. Taking the measured distance between snap-on and snap-off points multiplied by the spring constant of the cantilever as the adhesion force, F_{adh}, an almost constant G_A or G_m is obtained. In the batch analysis, F_{adh} is treated as a fit parameter to increase accuracy.

per unit area G_A, and hence the conductance per molecule G_m, the tip-SAM contact area has to be determined and combined with the SAM density of one molecule per $0.214\,\text{nm}^2$.[172] A universal contact model for the layered system SAM (of thickness t_{SAM}) on gold is not available. However, for sufficiently large contact radii, where the contact radius is larger than the film thickness ($r_{\text{contact}} > t_{\text{SAM}}$), a thin-film compression model, as described in Chap. 2.1.1, applies to this system[38, 37] and was used to calculate the contact area versus load. One of the largest sources of error in the data analysis is the uncertainty of the tip-surface adhesion force. Within the limits of the statistical scattering of the measured adhesion force, the adhesion force was used as a fit parameter to optimize for a constant thermal conductance per unit area for any given SAM. As depicted in Fig. 5.6 c), curves of constant G_A are obtained, except at low loads where the model assumptions do not apply. Because of the large thermal conductivity of gold and silicon, the spreading resistances in tip and substrate contribute only a negligible offset (see Chap. 5.2.2).

Figure 5.7 shows the thermal conductance as a function of the chain length N. Short-chain samples ($N \leq 4$) are significantly more conductive than the longer-chain samples with a maximum conductance at $N = 4$. This behavior was predicted by Segal $et\ al.$[162] and matches the data well, whereas other predictions[164, 170] appear to be incompatible with the observed conductances. Despite the scatter in the data, some significant conclusions can be drawn by comparing the results to existing simulations and data.

An overall decaying trend can be fitted empirically with $G = \dot{Q}/\Delta T = k/t_{\text{SAM}} \propto N^{\alpha-1}$, resulting in an exponent $\alpha = 0.59 \pm 1$ (from least squares). Since it has been argued that for very short chains different effects may play a role, discussing the data by separating it into a decaying and an ascending trend reveals a different picture. A diverging conductance for the decaying trend ($N = 4$ to 18) with $G = \dot{Q}/\Delta T = k/t_{\text{SAM}} \propto N^{\alpha-1}$ can be fitted, resulting in an exponent $\alpha = 0.38 \pm 0.2$ (from least squares) that is not compatible with either constant ($\alpha = 1$, as in fully ballistic transport) or

Fourier-like ($\alpha = 0$, as in diffusive transport) conductance. At first sight, this may be indicative of a situation in which transport is neither fully ballistic nor diffusive. A decreasing conductance with chain length between these extremes is, in principle, predicted in a quasi-ballistic transport regime with a phonon mean free path Λ on the same order as the chain length,[191] following $G \propto (N+\Lambda)^{-1}$. A fit with Λ expressed in multiplies of CH_2-units of the molecular backbone to the data yields $\Lambda \approx 5$; however, such a short mean free path is in striking contrast to a mean free path of 40 nm estimated from the thermal conductance measured using crystals of alkane chains (polyethylene) by Shen *et al.*[177] Therefore, a more plausible scenario is to assume ballistic transport through the carbon chains.[157, 158, 162, 170] Within its uncertainty, the data supports such a picture that leads to a constant conductance at least for $N \geq 15$, in accordance with the notion of interface-governed ballistic thermal transport.[166, 168, 170] Assuming ballistic transport through the molecules, the increase of conductance with decreasing chain length implies an increasing number of transmission channels or an increased heat-flux through the remaining channels. However, the number of vibronic states increases in proportion to N, with an approximately constant density of states.[162, 170] Therefore, although ballistic transport is expected here, also other mechanisms may play a role.

According to Patton and Geller,[192] the mechanical coupling between two contacts works directly. Instead of describing the ballistic transport as a number of steps, transmitting phonons into the molecule, propagation through the molecule and transmitting out of the molecule, the molecule serves as a mechanical linker. By neglecting the internal vibrational modes of the molecules, the coupling between the silicon tip and gold through the SAM can be modeled with single springs using the SAM's Young's modulus (40 ± 20 GPa). With a suitable offset of 8 pW K^{-1} accounting for ballistic low-frequency phonons, this allows a quantitative description of the data for $N = 4$ to 18. Still, the model predicts an even better mechanical coupling for $N = 2$ or 3, with at least one order of magnitude enhanced

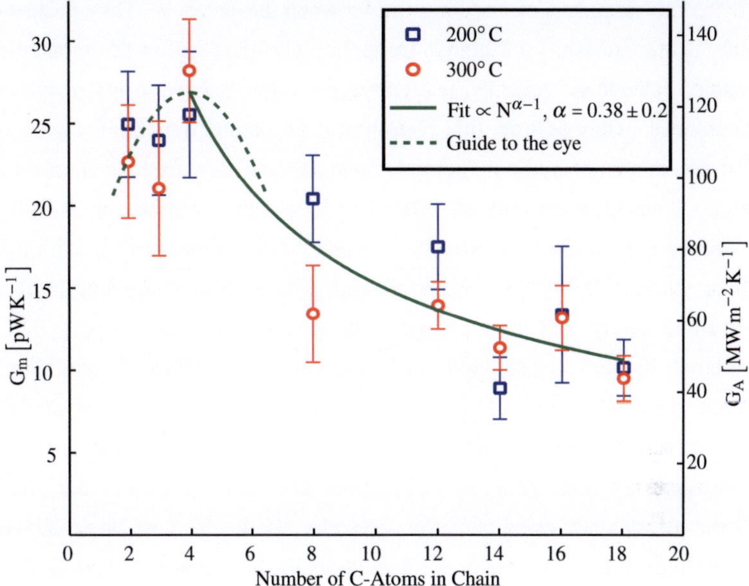

Figure 5.7.: Molecular thermal conductance as a function of chain length. The thermal conductance can be fitted with an $\alpha = 0.38 \pm 0.2$ for chain lengths longer than 4 carbon atoms in the chain. For shorter chains, the reduction of vibrational modes in the chain decreases the thermal conductance again, revealing a peak conductance around 4 carbon atoms per chain within the limits and errors of the experiment. A systematic error offsetting all data by $\pm 42\%$ is excluded.

conductance between $N = 2$ and $N = 4$. Therefore, it appears likely that the internal vibrational properties of the SAM layer play an increasing role in the increased phonon transmission for a decreasing chain length in the temperature range of applied here, especially for short alkane chains.

The opening or enhancement of transmission channels as required by the data under the assumption of ballistic transport may originate from phonon interference effects within the alkane chains. Simulation data[164] however, is available only for a few chain lengths and does not compare well with these measurements. Based on a very different physical approach, Segal *et al.*[162] also propose a direct coupling between the contacts. They propose phonon transmission via higher-frequency phonons within the molecule coupled to the low-frequency, heat-carrying phonons using classical quantum mechanics. They assume this coupling to weaken exponentially with N. The exponential decrease in thermal conductance with N for high-frequency phonons could potentially be related to localization effects due to fully delocalized phonons with specific maximum frequencies scaling with the chain length.[158] With full delocalization, fewer high-frequency modes can significantly contribute to thermal conductance for increasing N. This explanation predicts a stagnation of conductance, which can be explained by the reduced number of phonon frequencies in extremely short chains. The measured extent of increased conductance is related in this sense to frequency-dependent phonon localization. This explanation also accounts for the stagnation of conductance observed for $N < 4$, which can be explained by the reduced number of phonon frequencies in extremely short chains. However, the error bars of the data are too large to allow a final conclusion on this point. Nevertheless, the behavior predicted by Segal *et al.*[162] matches the data well, including the observation of a stagnating conductance for the shortest chains ($N < 4$).

5.2.6. Error Discussion

Possible sources of errors can be separated into systematic errors giving offsets to all the data in the same manner, and random scatter. The systematic errors affecting all data points in the same manner result in an uncertainty of the axes in Fig. 5.7. Errors, which can influence data points individually, are included in the error bars of the individual data points. The sources of systematic errors are the uncertainty of the heater/sensor calibration of 30 %,[12] uncertainties of the applied force due to the spring constant calibration of the cantilever of 30 % (from scanning electron micrographs of the cantilever[12]), the uncertainty of the tip radius determined by SEM of

10 % (see Fig. 3.9 c)), as well as the not very well defined Young's modulus for the SAM which is estimated to be about 50 % around a value of 40 GPa (see above). Using Gaussian error propagation, a total systematic error of ±42.5 % has to be applied to the vertical axes of Fig. 5.7., but not to the relative dependence of the data points. Individual scattering affecting single data points is included in the error bars and calculated from up to 3600 thermal contact curves per sample and heater temperature. These are statistics in the fitted curve (e.g. the adhesion force), and the unlikely inhomogeneity in the SAM itself and from SAM to SAM, which are taken into account by a variation of the Young's modulus by 5 % as well as experimental scatter. To sum up all these sources into one error bar, the biggest possible error caused by all those sources is estimated for each data point. For the error bars a Gaussian error propagation is not used but instead the maximum error is calculated to give a conservative estimation of the chain-length dependence of the thermal conductance. The measured thermal conductances and statistical errors for all heater / sensor temperatures and molecules are summarized in Tab. 5.1.

Molecule	G_m at 200°C $\left[\mathrm{pW\,K^{-1}}\right]$	σ $\left[\mathrm{pW\,K^{-1}}\right]$	G_m at 300°C $\left[\mathrm{pW\,K^{-1}}\right]$	σ $\left[\mathrm{pW\,K^{-1}}\right]$
$CH_3 - (CH_2)_1 - SH$	24.95	3.20	22.68	3.44
$CH_3 - (CH_2)_2 - SH$	23.97	3.34	21.04	4.09
$CH_3 - (CH_2)_3 - SH$	25.53	3.82	28.19	3.14
$CH_3 - (CH_2)_7 - SH$	20.41	2.65	13.41	1.95
$CH_3 - (CH_2)_{11} - SH$	17.49	2.59	13.92	1.44
$CH_3 - (CH_2)_{13} - SH$	8.89	1.88	11.35	1.35
$CH_3 - (CH_2)_{15} - SH$	13.37	4.15	13.16	1.97
$CH_3 - (CH_2)_{17} - SH$	10.14	1.71	9.487	1.28

Table 5.1.: Thermal conductance of the alkane linker as measured for 200°C and 300°C heater temperature.

5.3. Conclusion and Outlook

In conclusion, thermoresistive sensing achieved sufficient sensitivity to quantify the thermal conductance of alkane-thiol SAMs as a function of chain length by applying a thermal force mapping technique to scanning thermal microscopy. Here, the heat transport is dominated by the interfaces between the tip and the SAM. This requires a precise description of the contact mechanics. By thermal force mapping, variations in the chain-length-dependent thermal conductance of a factor of 3 can be observed for short alkane chains, with a maximal conductance for 4 methylene units per chain. The decaying trend for $N \geq 4$ neither satisfies requirements of diffusive (Fourier-like) transport nor exhibits constant conductance as previously predicted. For alkane chains longer than four methylenes, a diverging conductance ($\alpha = 0.38$) is observed, similar to what is expected for very long chains. However, the increase of the thermal conductance for a chain length of four carbon atoms and the decrease to a quasi-constant conductance for chain lengths of up to 20 carbon atoms are also predicted for short chains in recent simulations and are potentially related to phonon localization effects. The high sensitivity to the interfaces between SAM and tip also requires chemically stable interfaces for quantitative analysis.

The proposed thermal force mapping technique has shown sufficient sensitivity and reliability for probing molecular conductance in the model system of linear alkane chains. This offers an experimental tool to probe various not fully understood molecular systems. Especially, changing the vibrational frequencies of the molecules by replacing the hydrogen bonded to the carbon atoms by heavier atoms is a simple modification to the examined molecules and allows to test the dominant heat transport path by changing the molecule vibrational modes. Additionally, scattering centers for phonons can be introduced into linear chains e.g. by arranging ether or aromatic linkers to connect two or more alkane chains. Understanding of the basic principles of heat transfer in such organic interface materials then enables the specific thermal design of devices down to the molecular level.

*A shorted version of this chapter has been accepted as the article "Length-Dependent Thermal Transport Along Molecular Chains". T. Meier, F. Menges, P. Nirmalraj, H. Hölscher, H. Riel and B. Gotsmann: Physical Review Letters **113**, 060801 (2014).*

6. Replication of Nano- and Microstructures on Curved Surfaces

6.1. Introduction

Micro- and nanostructured surfaces are widely used in modern microsystem applications as they enhance surface effects (e.g. adhesion)[193] or enable new applications.[194, 195] With efficient replication techniques, micro- and nanostructured polymer devices are also well represented in the consumer mass market. The development of new devices for optical applications or lab-on-a-chip systems are often time consuming and, therefore, cost-intensive. Replication of polymer micro- and nanostructures with metallic masters is a common way to reduce these costs per piece.[67, 196] The fabrication of these metallic masters is a very complex process and the lifetime of such master structures is limited to a maximum replication cycles. Recent developments, however, had shown that even polymeric masters can be used as molds for polymer replication.[197] As the fabrication methods for fabricating the master structures are limited to flat substrates, this limits the replication techniques using micro- and nano-structured molds as well as the 2.5 dimensional features because of the demolding process. The demolding can only be done perpendicular to the mold and undercuts in the structures cannot be demolded. This technique of replicating structures on the surface of a device is well established. However, the integration of such structured surfaces with specific functionality into microsystems is still not easy for most applications and can be done in multistep processes of hot embossing, microthermoforming and bonding.[198] Microthermoforming is used to transform micro- and nanostructured thin foils from a planar shape into a three dimensional curved micro- and nanostructured surface. For rigid and bulk materials, microthermoforming however is not suitable.[199]

To replicate three dimensional micro and nanostructures with undercuts on curved surfaces of rigid and bulk materials, a mold with new feature characteristics is introduced by utilizing shape memory materials for its fabrication. Specifically, using shape memory polymers as a mold material, the mold does not only provide the shape to be replicated (like for 2.5 dimensional structures) but has also an additional functionality as it can change its shape. Shape memory polymers are advantageous for such applications compared to other materials like shape memory alloys, as their shape changing capability is rather large[76,77] and they can be processed like other polymer. Therefore, they can be processed by well established micro- and nanoreplication techniques like hot embossing and can also be used as mold materials for this techniques. Shape memory polymers are an especially interesting group of mechanical active materials, because they are capable of single, dual or multiple shape changes[72,73,74] on the nanoscale activated by external triggers (e.g. heat or light). Commercially available polymers like Tecoflex ® EG 72D and Tecoplast® TP 470 are thermally activated shape memory polymers of high interest for this application. Both are capable of recovering a predefined macroscopic and microscopic shape after they have been deformed into different temporary shapes.[76,77,78,79,80,200,201,202] Using the shape memory effect in a mold can help to reduce the wear of the mold due to temporary deformations of the mold micro- and nanostructures due to high loads on this structures and enables new features in designing microsystems with curved surfaces or in bulk materials patterned with microstructures. During the development and optimization of a replication process using shape memory polymer molds, large area atomic force microscopy is advantageous compared to other surface characterization methods. The large area scanning allows the analysis of the mold's features on a large area with a high resolution. Additionally, as the polymers used here are not conductive, other characterization methods like scanning electron microscopy adverse sample preparation like deposition of gold on the polymeric molds. Such invasive characterization methods prohibit a further use of the polymeric mold and can, therefore, only be used for final investigation of the mold, but not for in-process screening. Optical methods, on the other hand, cannot provide the resolution required for the characterization of micro- and nanostructures.

6.2. Characterization of the Shape Memory Effect of Tecoflex® EG 72D

Shape memory materials form a technologically important class of materials for actuators, sensors and medical applications.[203,204,205] The subclass of shape memory polymers allows to engineer material properties like the trigger of a shape transition, glass temperature or maximum deformation strain for specific applications. In general they are much cheaper and easier to process than classical shape memory alloys.[57,79,200] To use a shape memory polymer for mold fabrication, two commercially available shape memory polymers were chosen. In order to fabricate stable and reliable molds out of shape memory polymers, two major properties have to be investigated. The mold must show a stable permanent structuring of small structural details and a precise control of the restoring from a temporary shape to the permanent shape. Even as there exist shape memory polymers capable of multiple shape changes[72,73,74] on laboratory scale, one way shape memory polymers like Tecoflex® EG 72D and Tecoplast® TP 470 are promising candidates for process development as they can be purchased in large quantities at a constant quality level.

To benefit from the shape memory effect for the mold fabrication, the material must enable the reliable switching between at least two distinguished states. As the polymer is designed to enable the switching on the macroscale while heated above the transition temperature T_{trans}, it is not clear if this switching between distinguished states can also be observed on the micro

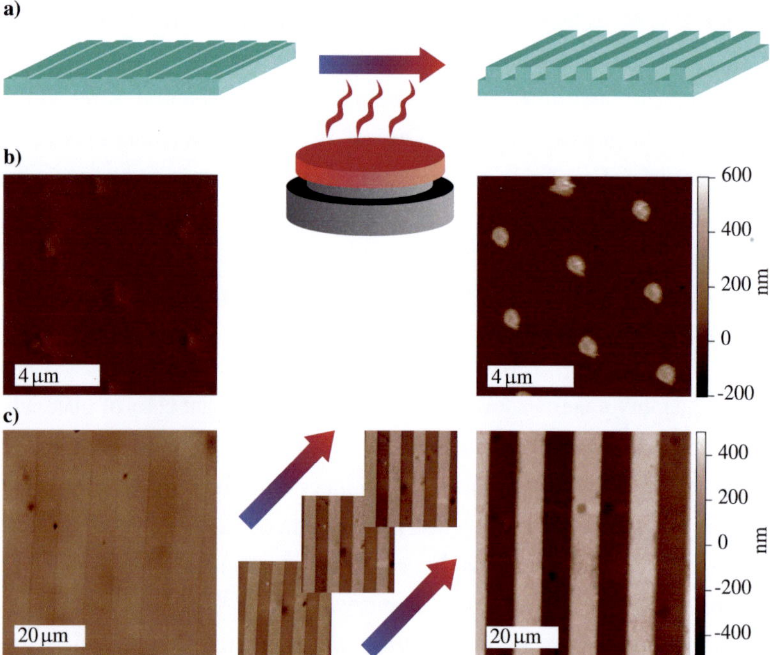

Figure 6.1.: Concept of replicated micro and sub-micro structures on demand fabricated out of shape memory polymers. Using a shape memory polymer, micro- and nano-structures are replicated by a first hot embossing step into the polymer device. In a second embossing step, a flat surface has been programmed temporarily. **a)** By heating the device above T_{trans}, the temporary structures vanish and the permanently programmed structures reappear. **b)** AFM images of freestanding pillars of 500 nm diameter and height before and after the recovery of the permanent shape. **c)** A permanently embossed optical phase grating monitored with the AFM over time while the permanent structure reappears.

and nanoscale. Structuring a permanent shape into the polymer by hot embossing is comparable with hot embossing of thermoplastic polymers, however, Tecoflex® EG 72D and Tecoplast® TP 470 have two glass transition temperatures which allow a permanent structuring above the higher transition temperature $T_{\text{g,ht}}$ and a temporary programming between the lower temperature $T_{g,lt}$ and the higher temperature $T_{g,ht}$. By hot embossing shape memory polymers, the programmed shape has to be fixated which is done by a complete relaxation of the polymer in order to form a stable grid within the material. This grid is stabilized by the interaction of netpoints within the polymeric structure. With smaller structure details, the number of grid netpoints within the features decreases which can alter the fixation of the permanent and temporal shape.

Typical applications of shape memory polymers, like for example heat shrinkage tubes for electric isolation, use the shape memory effect on the macroscale with features down to the millimeter scale.[78,205] As shown in Fig. 6.1, a reliable and stable fixation of both the temporal shape and permanent structure is indeed possible for structure details down to the sub-micron range. As the permanent shape can be structured by hot embossing above $T_{g,ht}$, the fixation of the temporary shape requires the right temperature for the transformation into the temporal shape as well a "freezing" of this shape. In general, the transformation into the temporary shape has to be done above T_{trans} but below $T_{g,ht}$ to keep the netpoints of the permanent shape. Like shown in Fig. 6.2, the transformation temperature is not a sharp edge above which the shape memory effect is triggered. The shape memory effect is also triggered below this temperature, but the restoring speed is small. For a stable fixation of the temporary shape of Tecoflex® EG 72D a cooling of the material from T_{trans} down to at least room temperature while holding the temporal shape is required, smaller structures even benefit from lower temperatures.

In order to quantify the shape change from a temporary shape to the permanent structure, an optical phase grating is hot embossed into the shape memory polymer Tecoflex® EG 72D. In a second step, a flat surface topography is programmed temporarily. The restoring process from the temporary shape to the permanent shape is then measured at different temperatures with an AFM. While the sample is heated to a constant temperature, the AFM measures the topography, which will change over time from the temporary

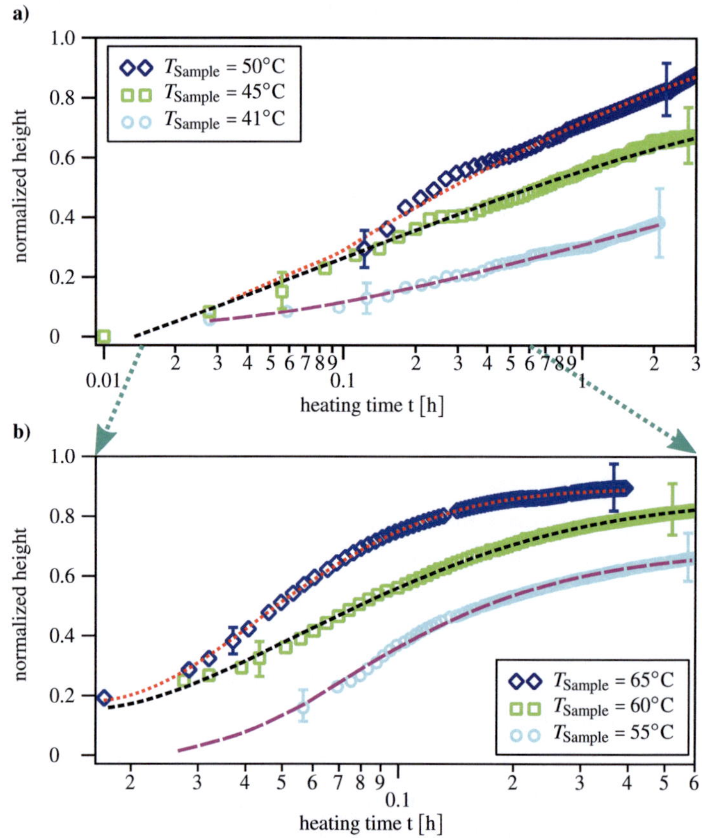

Figure 6.2.: Recovery rate of the permanent structures in Tecoflex® EG 72D over time for different temperatures. The recovery rate was measured by a histogram analysis of persistently captured AFM images of a line grating as shown in Fig. 6.1 c). The rated transition temperature of this polymer is $T_{\text{trans,TFX}} = 55\,°C$. Therefore the recovery of the permanent shape should also occur around this temperature. A full recovery can be observed for long heating times. **a)** The recovery rate for temperatures below the transition temperature. The recovery rate is strongly temperature dependent and very slow. However, the recovery speed increases with increasing temperature. **b)** For temperatures around and above $T_{\text{trans,TFX}}$, the growing of the structures follows again a logarithmic growing law (fits as dashed lines in the graph) also a increasing speed with increasing temperature. Fit parameters can be found in Tab. 6.1.

shape into the permanent shape. From picture to picture, the topography of the sample changes and can be used as a measure for the shape change. In fact, the topography will change from scan line to scan line. This change, however, is included into the errorbars of the individual data points in Fig. 6.2 as the median grow is used as the shape change of the picture. This procedure was repeated for six different temperatures between $41°C$ and $65°C$ which is around $T_{\text{trans,TFX}} = 55°C$. The structure height is then normalized to the height of the permanent shape for fitting the data with a normal logarithmic growing law. Normalizing the height enables to use the obtained normalized height as a function of time as a factor to estimate the state of transformation for any other structure. For the growing, a normal logarithmic growing law is assumed. The growing can be described by the equation:

$$h(t) = h_0 + A \cdot \exp\left(-\left(\frac{\ln(t-t_0)}{\tau}\right)^2\right)$$

(6.1)

Here, the normalized height $h(t)/h_0$ of the structures varies between 0 and 1 as a function of time t. As shown in Fig. 6.2, the recovery process can be fitted with this approach well for all temperatures measured here. Interesting at this point is, that a slow recovery can also be observed below the rated switching temperature $T_{\text{trans,TFX}}$. This behavior can be explained by the molecular structure of the polymer as the glass transition temperature of a polymer is a function of chain length and degree of polymerization which can vary around a mean value. As the glass transition temperature $T_{\text{g,lt}}$ especially of the soft segments is defining the transition temperature T_{trans}, the switching of the micro and nanostructures also occurs in a temperature regime around T_{trans}.

The fitting parameter of interest for determining the transition temperature range is the time constant τ which defines the recovery time. In Tab. 6.1, the time constants of the recovery time are given. The recovery of the structures also starts very slowly at temperatures well bellow the rated

T_{Sample}	$\tau\,[\ln(h)]$	$\sigma_\tau\,[\ln(h)]$
41°C	6.4956	1.14
45°C	4.8195	0.011
50°C	2.5047	0.4430
55°C	2.0692	0.0143
60°C	1.6030	0.0492
65°C	1.4644	0.0152

Table 6.1.: The restoring of the permanently programmed shape follows a normal logarithmic growing law and was fitted (using least squares) to Eq. (6.1). The time constant $\tau \pm \sigma_\tau$ is a measure how fast the permanent shape of Tecoflex® EG 72D is restored for a temperature regime between 41°C and 65°C.

transition temperature T_{trans} = 55°C. However, the time constant of the recovery is highly temperature dependent. As expected, it decreases with increasing temperature. By using the time constant to describe the recovery process, it can be concluded that a stable temporary shape is characterized by a very high time constant. The increase of recovery speed with respect to the temperature is also represented by the time constant, as increasing temperatures reduce the time constant. As the time constant is reduced with respect to temperature rises significantly for temperature rises below T_{trans}, the time constant is less temperature dependent for temperatures above T_{trans}. The recovery speed can also be controlled by adjusting the temperature in the desired range.

The shape memory effect can also be triggered locally by a locally confined heating of the polymer. As seen before, the restoring of the permanent shape is both a function of temperature and heating time. By confining the heat in a very narrow region in the sample, the restoring of the permanent structures is only triggered in this region. By including 25 µm thin heater wires into the Tecoflex® EG 72D devices, a localized electrical heating is achieved.[206] The wire was embedded during the permanent structuring of a 60 µm thick foil on the unstructured side of the foil. Optical microscopy of

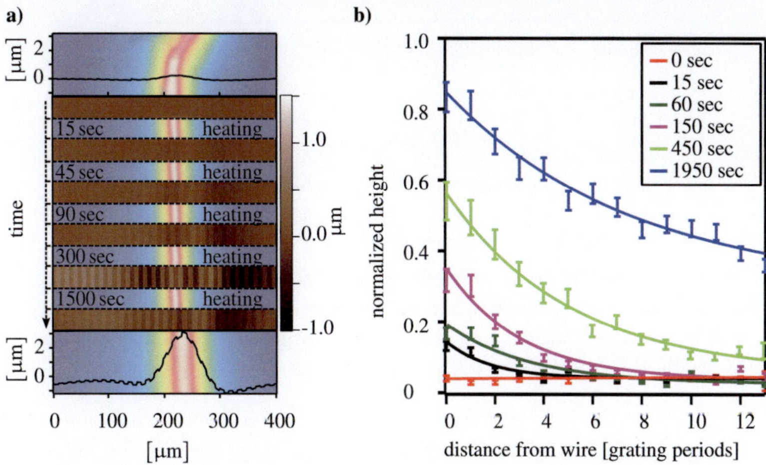

Figure 6.3.: Local recovery of the permanently programmed structures in Tecoflex® EG 72D by heating with an embedded wire. A phase grating has been fabricated like in the experiments before while an embedded 25 µm thick heater wire was introduced in the shape memory polymer. The embedding of the wire and fabrication of the of the sample was done by Senta Schauer.[206] **a)** The wire was heated with an electrical power of 10 W while the surface temperature of the sample was monitored with an IR-Camera. After subsequent heating, the structure details investigated on large area AFM images revealing a local recovery of the surface structures. **b)** The normalized structure height as a function of the grating periods on the sample. The grating has a periodicity of $\Lambda = 16$ µm.

cross-sections show the wire being completely embedded in the foil. The distance from structured surface of the foil is, therefore, estimated between $30 - 40$ µm. After subsequent heating by the wire, the structure height is measured with the AFM and subsequently analyzed as a function of distance from the wire. The position of the wire can be identified in the AFM images by Gaussian-like surface profile caused by the higher stiffness of the wire and, therefore, less deformation of the wire during the embossing and programming the temporary shape.

Fitting of the height of each individual structure as a function of distance from the wire can be done using an exponential decay. Note that only the

t_{heating}	λ [grating periods]	σ_λ [grating periods]
15 s	1.841	0.435
60 s	2.418	0.882
150 s	3.077	0.468
450 s	5.080	0.683
1950 s	7.647	1.81

Table 6.2.: The local restoring of the permanently programmed shape follows an exponential decay function and was fitted (using least squares) to Eq. (6.2). The half-value of length $\lambda \pm \sigma_\lambda$ is a measure how locally the permanent shape of Tecoflex® EG 72D is restored.

height of the features is fitted, not the overall surface shape due to the wire profile.

$$h(x) = h_0 + A \cdot \exp\left(-\frac{x}{\lambda}\right) \qquad (6.2)$$

Here, the normalized height $h(x)/h_0$ of the structures varies between 1 and 0 as a function of the distance from the wire in multiples of the grating period of 16 μm. The parameter λ can be interpreted as a half-value of length and characterizes the local confinement of the restoring process.

In Tab. 6.2, the local confinement of the restoring is given for different heating times. Combining both, the time constant τ and local confinement λ allows to control recovery speed and local recovery by local heating of the polymer. More important, the recovery process can also be stopped and restarted for the desired application as a material for the mold fabrication. The structure height was analyzed from as a function of the distance from the wire edge. When assuming the temperature distribution in the polymer as a function of the distance from the wire as a hot spot, heat diffusion in the polymer can be described by

$$\frac{\partial^2 T(r)}{\partial r^2} - A(T(r) - T_0) = 0, \qquad (6.3)$$

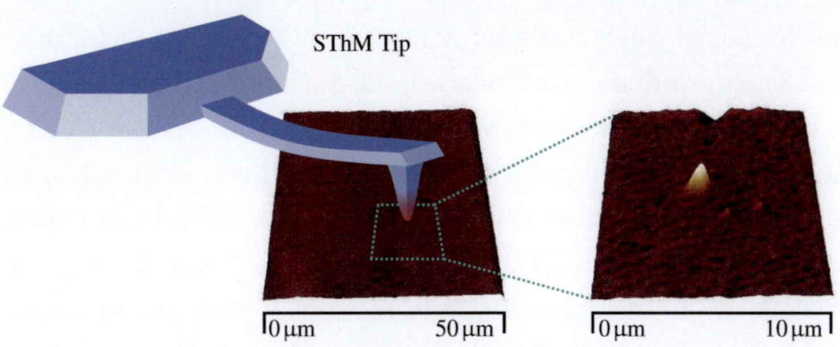

Figure 6.4.: Concept of "Nanostructures on Demand". The shape memory polymer foil was temporary programmed to be thinner than the permanent shape. By a locally confined heating using a heated probe, a pin has grown out of the foil. By improving force an temperature control, this can be used as a lithography method for fabricating nanostructures on demand.

including a material constant A which includes the thermal conductivity of the polymer, interface resistances between wire and polymer by assuming an unidirectional thermal transport in the polymer.[207] For the boundary condition of $T(r=0) = T_0$, the temperature profile in the material follows:

$$T(r) = T_0 \exp\left(-\frac{r}{\sqrt{A}}\right) \qquad (6.4)$$

As found before in Fig. 6.2, the recovery time constant $\tau(T)$ is a (non-linear) function of the temperature. As the temperature on the structured surface is a function of the distance to the wire, also the recovery time constant on the surface is a function of the distance from the wire. The restoring speed on the surface is, therefore, a function of $\tau(T(r))$. However, as there are many parameters in the experiments carried out, which cannot be controlled accurately, like for example the distance from the wire, additional experiments have to be conducted. Therefore, for the experiments presented here, an exponential decay for fitting the distance from the wire dependent restoring is a good approximation.

To define the position of the hot spot more precisely, a heated tip of a scanning thermal microscope can be used as shown in Fig. 6.4 and can, therefore, allow to address single structural features for restoring to the permanent shape. These experiments are conducted with commercially available heated tips in ambient conditions. However, those tips are heated with a micro-patterned printed circuit on the silicon cantilever. For a confined heating with a precise force control as needed for such an experiment, this probe design has two major drawbacks. First, the tip radius is very large and in the range of a few 100 nm. More seriously, however, is the second drawback. The thin metal film near the tip apex used to heat the probe also introduces a pre-bending due to a different thermal expansion than the silicon cantilever. This pre-bending and thermal expansion of the cantilever, which is tempera-ture dependent, can offset the tip position on the sample and alter a precise force control. Using more reliable heated probes can, therefore, significantly improve this concept of on demand addressable nanostructures.

6.3. Shape Memory Mold Fabrication and Thermoshaping

In order to use the shape memory polymers for the fabrication of molds with new functionality, such molds have to be reliable in classical replication methods. Therefore, in a first step 2.5 dimensional shape memory molds are fabricated by a one step hot embossing process using a master mold fabri-cated out of nickel. The shape memory polymer must show a permanently stable shape. This can be assured for Tecoflex$^{®}$ EG 72D by structuring the material above $T_{g,ht} = 150\,°C$ and for Tecoplast$^{®}$ TP 470 above $T_{g,ht} = 190\,°C$, respectively. For the fixation of the mold, the embossing force of 200 kN for both polymers is hold during the cooling of the polymer to room temperature. For classical replication techniques of 2.5 dimensional microstructures, the shape memory device is ready to be used as a mold. In Fig. 6.5 and Fig. 6.7, examples of replication by hot embossing and casting using molds fabricated out of shape memory polymers are given.

Shape memory polymer molds show the same functionality of previously presented polymeric molds fabricated out of high performance polymers.[197]

However, deformations of the mold due to embossing forces only deform the low temperature soft phase of the shape memory polymer. Using the shape memory effect of the material allows the restoring of the permanent shape to heal mold deformations.

However, not only the healing of deformation is an advantage of shape memory polymer molds. Additionally, a new functionality to replication with molds is introduced, as the shape memory effect of the material can be used for other purposes than just healing deformations. For this new functionality, the "thermoshaping" process is introduced, which uses the temporary shape of the material. For this step, a ready fabricated shape memory mold is reprogrammed in a temporary shape. The temporary shape can follow any three dimensional surface - the permanently flat mold is adapted to curved surfaces, folded around angles or partially elongated. If the thermoshaped mold is filled with a new polymer, e.g. by casting, the shape memory mold can demold itself or its micro- and nanostructures using the shape memory effect. Thereby, the transformation from the temporary to the permanent programmed shape demolds the casted polymeric device to a point, where undercuts limit a classical demolding.

Thermoshaping introduces a new class of molds for embossing, nanoimprint and casting, manufactured from shape memory polymers. These polymer molds can be easily fabricated and show self-healing properties due to the shape memory effect of the mold material. Furthermore, those molds allow to replicate three-dimensional micro- and nanostructures. To demonstrate the concept of a self-demolding mold, a solid epoxy rod with microstructures on its surface and a helical microchannel with internal microstructures are shown in Fig. 6.8 and Fig. 6.9.

6.4. Replication with Shape Memory Polymer Molds

If the material of a mold is replaced by a shape memory polymer, the mold must still be capable to be used in well established replication processes.

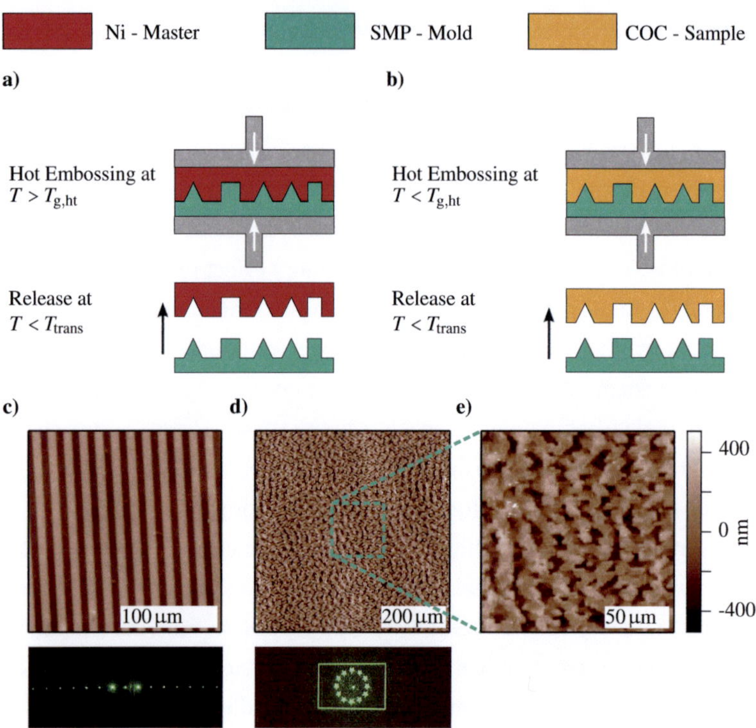

Figure 6.5.: Devices which need optical grade surface quality can be replicated by hot embossing.[101] However, they require defect-free molds with the same high surface quality. A Tecoplast® TP 470 mold is used to emboss optical phase gratings into COC Topas 8007 at 115°C embossing temperature and 7000 N (3 kPa) embossing force. **a)** A simple line grating replicated in COC Topas 8007. The AFM topography image at the top as well as the interference pattern on the bottom reveals the high surface quality. **b)** With a more complex geometry, a grating showing the Fourier-transformed European flag with more complex topographic features and a periodicity of 500 μm. The large area scan reveals the defect free replication on these length scales. The corresponding interference pattern at the bottom confirms the high replication quality of microstructures on large areas. **c)** The zoomed area of the European flag revealing its 16 different levels in topography. The gratings have been fabricated on a single chip of 35 mm x 75 mm with multiple optical phasegratings. For the fabrication of optical structures, the optical surface quality of the mold can be restored with the help of the shape-memory effect right before every embossing step.

Hot embossing is such a well established and a very flexible method for microreplication and can also be used for the shape memory mold fabrication. However, shape memory molds can also be used for this replication method. While embossing with the shape memory mold, the embossing temperature T_{emboss} has to stay below $T_{\mathrm{g,ht}}$ to keep the permanent programmed shape. If $T_{\mathrm{emboss}} > T_{\mathrm{g,ht}}$, the shape memory mold gets reprogrammed. As the Youngs-modulus of the shape memory mold as well as the Youngs-modulus of the polymer to be embossed is strongly temperature dependent, the ideal embossing temperature using a SMP-mold is above $T_{\mathrm{g,Polymer}}$ and below $T_{\mathrm{g,ht}}$ at the minimal Youngs-modulus ratio $E_{\mathrm{Polymer}}(T)/E_{\mathrm{Mold}}(T)$. As mold and polymer get softer and softer at higher temperatures, the best embossing results are obtained while heating the polymer to $T_{\mathrm{emboss}} + \Delta T$ and the mold to $T_{\mathrm{emboss}} - \Delta T$ before starting the embossing. Depending on the heat capacity of the mold and the polymer, the ΔT must not be symmetrical around T_{emboss}. In this way a temperature gradient is introduced and the comparable soft polymer can be embossed with a relatively hard mold while the system starts to equilibrate. The optical structures shown in Fig. 6.5 are embossed into Cyclic Olefin Copolymer (COC) Topas 8007 with a glass transition temperature of $T_{\mathrm{g,COC}} = 75\,^{\circ}\mathrm{C}$. The mold has been fabricated out of Tecoplast® TP 470. The optimal ratio of $E_{\mathrm{Polymer}}(T)/E_{\mathrm{Mold}}(T)$ is experimentally found at $T_{\mathrm{emboss}} = 115\,^{\circ}\mathrm{C}$. To support the embossing by a relatively stiff mold pressed into a relatively soft polymer, a temperature gradient of $\Delta T = 45\,^{\circ}\mathrm{C}$ was used. The embossing was done with an embossing pressure of $3\,\mathrm{kPa}$ until mold and polymer thermally equilibrate before it was enhanced to $5\,\mathrm{kPa}$. Afterwards, mold and polymer were cooled down to $40\,^{\circ}\mathrm{C}$ and demolded.

The stability of a SMP-mold for hot embossing has been studied with the structures shown in Fig. 6.5 c). As a SMP-mold can be resetted to its initially programmed shape, the lifetime is not limited by the number of embossing cycles the mold can withstand without being deformed by the embossing forces. After 10 embossing cycles without resetting, the optical function of the replicated gratings was still sufficient. Compared to

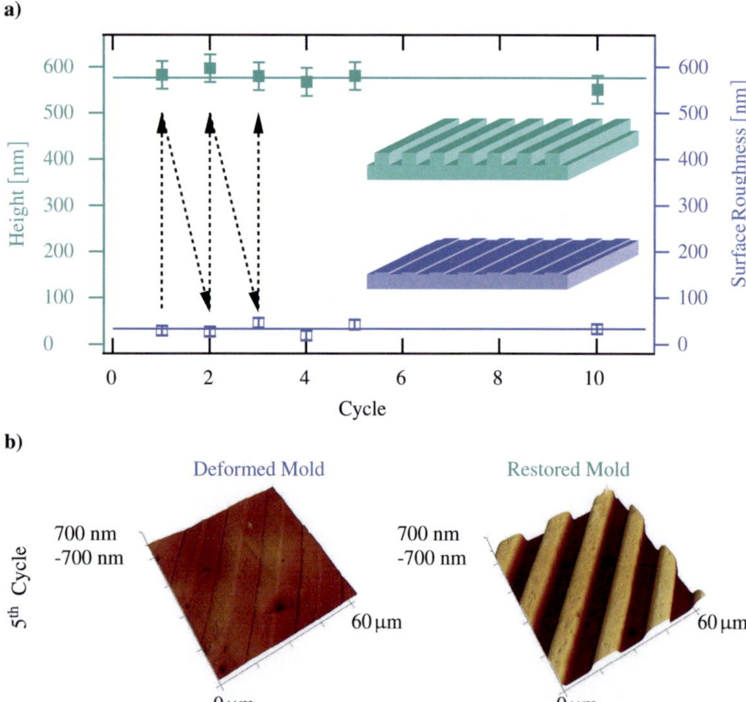

Figure 6.6.: Demonstration of reproducible cycling between deformed and permanent mold shape. **a)** Ten test cycle between the temporary and the permanent shape are analyzed in this graph. It compares the height of the permanent shape (an optical grating with a nominal height of 600 nm) with the surface roughness of the damaged mold. To damage the Tecoflex® EG 72D mold, it was flattened with a temperature of $T_{\text{trans}} < T < T_{\text{g,ht}}$ and a high pressure of 6 GPa between two silicon wafers. After cooling the sample to room temperature its topography was imaged with an atomic force microscope. Subsequent heating to T_{trans} recovered the permanent surface structure which was again analyzed by atomic force microscopy. **b)** Atomic force microscopy images of the deformed and restored shape of the Tecoflex® EG 72D mold during the 5th test cycle.

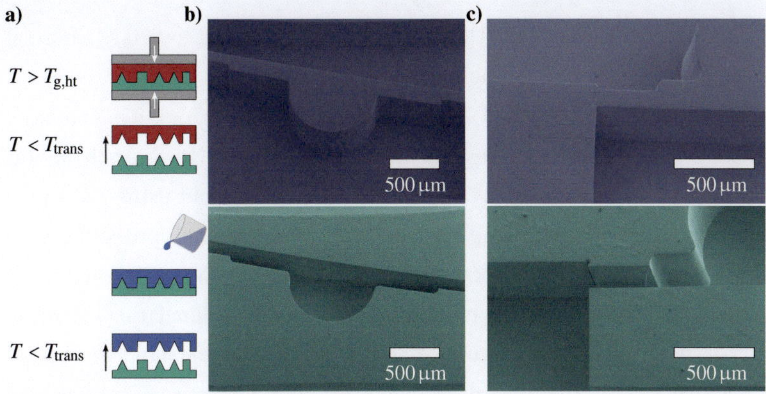

Figure 6.7.: Casting of high aspect ratio PDMS structures into a Tecoflex® EG 72D shape memory mold **a)** The mold is fabricated by a hot embossing step programming the structures permanently into the mold. In **b)** and **c)** SEM micrographs depicting the replicated structures in PDMS (top micrographs) and the master structures of the Tecoflex® EG 72D mold (lower micrographs). Aspect ratios of up to seven were realized with structure sizes in the micron range. The Tecoflex® EG 72D mold in the lower micrographs has been run through four resetting cycles (see Fig. 6.6) before casting of the PDMS structures above and capturing the SEM micrographs.

previously presented polymeric molds fabricated out of high performance polymers,[197] the SMP-molds used for hot embossing perform comparable even without the use of the shape memory effect. If resetted to the initial programming, the mold can withstand the same number of embossing cycles again. Especially for optically functional surfaces, the resetting provides a huge benefit concerning a long term stable surface quality of the mold.

For microfluidic applications, casting of soft-elastomers like polydimethylsiloxane is one of the standard fabrication techniques in academia for lab-on-chip systems.[56] For those applications, materials with well know behavior (to fluids, cells, sealing of channels...) is crucial as well as the ability of the fabrication of high aspect ratio structures. In Fig. 6.7 free standing structures and reservoirs with aspect ratios up to 7 were casted with a Tecoflex® EG 72D mold. The Tecoflex® EG 72D mold is used as a normal master for casting and can be reused for multiple parts.

At classical replication of 2.5 dimensional structures, the shape memory effect is not crucial for successfully replicated devices. However, the molds lifetime can be extended by resetting a deformed mold using the shape memory effect. For investigating the long term resetting properties of a deformed SMP-mold, deformation and resetting cycles of a mold were performed. The results are summarized in Fig. 6.6. By simulating enhanced embossing forces and loads to the features of the mold, the mold was heated up above T_{trans} to a temperature below $T_{\text{g,ht}}$ and a increased normal load on the functional features of the mold was applied. Those parameters are off scale compared to the standard embossing parameters as the embossing temperature $T_{\text{emboss}} < T_{\text{trans}}$ is lower and the applied normal pressure of 6 GPa is 6 orders of magnitude higher than the pressure of 3 kPa used for embossing the structures shown in Fig. 6.5. This high pressure was reached while the mold was pressed between two stainless steel plates instead of pressing a polymer into its cavities. At this pressure, the stainless steel is also deformed by the mold. Therefore, this scenario is probably exceeding typical pressures during hot embossing. As the applied pressure was extremely high compared to normal embossing parameters and the stainless steel is extremely hard compared to normal polymers, the structural features of the mold get flattened to less than 1/10 of its original feature height. After unloading the pressure from the mold, the features were imaged with an atomic force microscope. Heating of the mold to $T = T_{\text{trans}}$ triggers the shape memory effect and reveals the initial programming of the mold. The deformation-restoring cycles of the mold were stopped after 10 cycles showing a long term durability of the shape memory mold even under these extreme conditions. Fig. 6.6 a) demonstrates the reproducibility of the switching between a deformed and permanent shape.

The reproducibility of the switching between temporary and permanent shape is a precondition for a successful self-demolding of thermoshaped

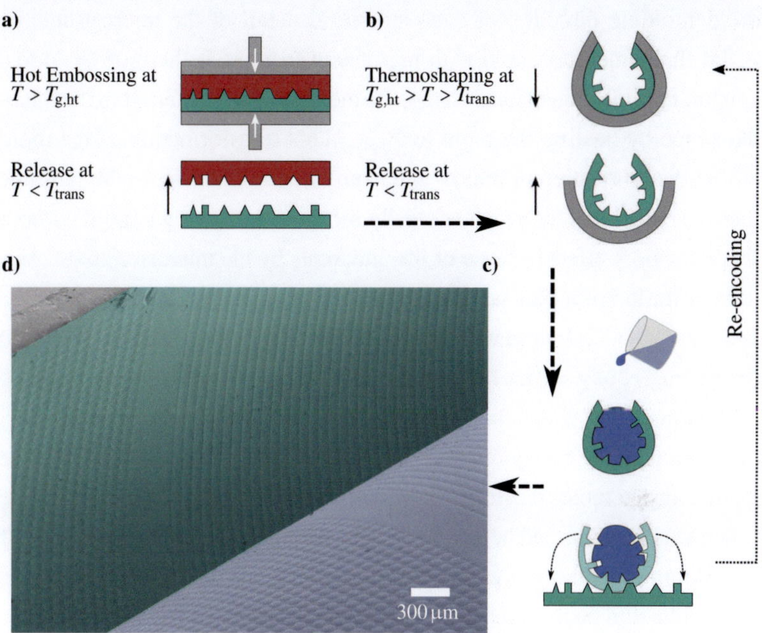

Figure 6.8.: Microlenses on a curved surface of an epoxy rod fabricated with a thermoshaped Tecoflex® EG 72D mold. **a)** In a first step, the microlenses were hot embossed as a permanent programming into Tecoflex® EG 72D. **b)** The flat mold is thermoshaped into a tube like temporary shape at a temperature above T_{trans}. **c)** The thermoshaped mold is filled with a two component epoxy. After curing, the solid epoxy rod is demolded by restoring the permanent shape of the mold. The mold peels itself of the epoxy and demold the microlenses in normal direction of the rod surface. **d)** The SEM micrograph shows the epoxy rod with microlenses on the outside. On the top, the partially demolded thermoshaped mold is visible with the epoxy rod on the lower right. The mold can be reused by reprogramming the temporary shape.

molds. At this point, self-demolding is defined as the mold's ability to use the shape memory effect of the transition between a temporary and a permanent shape for demolding. This can be done in two ways. In the first example, the temporary shape is programmed by thermoshaping on a macroscopic scale. The shape memory effect is used to provide the demolding force along the demolding direction of each structural detail of the microstructures. A flat shape memory mold with permanent features is thermoshaped to a temporary three dimensional shape, the mold can be transferred to the initial flat shape by heating the mold to T_{trans}. This transformation of the mold allows the fabrication of micro- and nano-structured devices with reusable thermoshaped molds, which normally requires destroying a mold without shape memory effect because of the undercuts by the microstructures. As a demonstration of a part fabricated with a thermoshaped mold, a tube-like Tecoflex® EG 72D mold with microlenses as functional features was used for casting epoxy. After the polymerization of the epoxy, the demolding of the epoxy rod was done by heating the mold to T_{trans}. For demolding, the mold with the epoxy rod was placed for 10 min in a oven at 60 °C. The thermoshaped mold transforms itself to the initially programmed flat shape which was programmed before the thermoshaping procedure. After this self demolding step, the epoxy rod could easily be taken out of the mold, while the mold could be reused for another thermoshaping step with a similar or completely different temporary shape.

The second feasible way to use the shape memory effect for self demolding is to program the microstructures temporarily into the shape memory polymer while the permanent shape is a flat surface. Restoring the permanent shape from the temporary shape makes the microstructures disappear. In this way, only the demolding of the micro and nanostructures can be done which becomes very practical for the fabrication of microfluidic devices. The shape memory mold can be used as a core defining the dimensions and surface structures of a microchannel. Microstructures in polymeric microfluidic devices can provide specific surface properties like wetting behavior[208] or protein adsorption.[198] However, patterning the sidewalls of microchannels is mostly done by structuring a surface which is used to cover a fluidic backbone. The structured surface is thereby connected by bonding techniques with the backbone enhancing the number of fabrication steps and the risk of leakage. Using a shape memory mold with

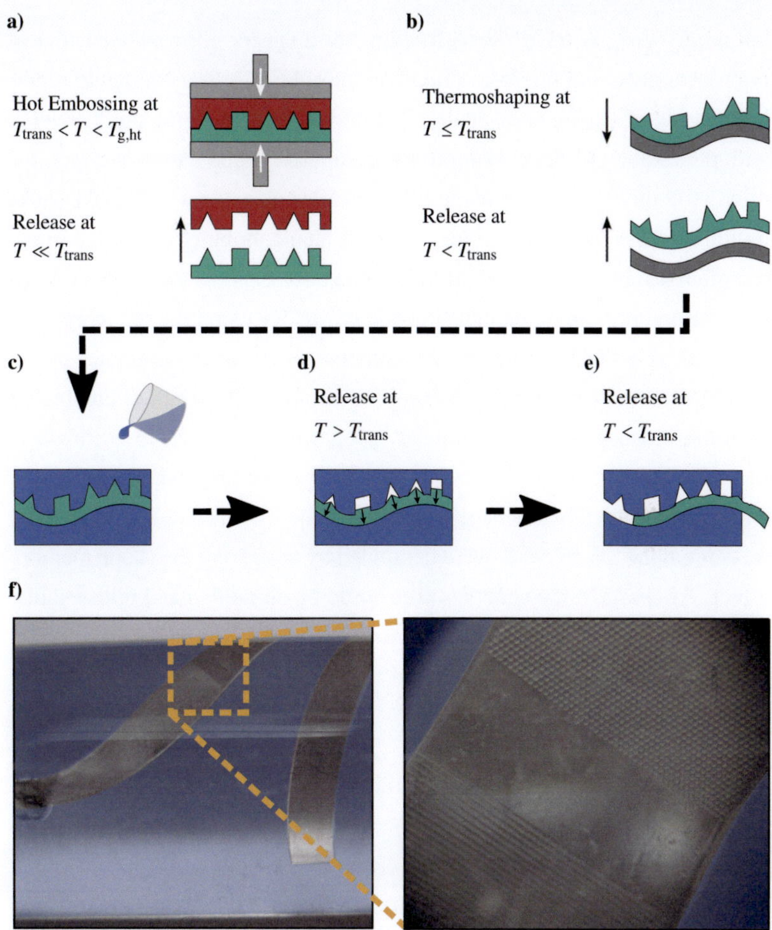

Figure 6.9.: Helical microchannel with microlenses on the channel walls fabricated by self-demolding shape memory molds. **a)** The microstructures are temporarily programmed into the shape memory material. **b)** The thermoshaping is done below T_{trans} to keep the temporary programmed microstructures on the mold. **c)** Again, PDMS is used as a casting polymer. The mold is used as a microstructured spacer for a later release of the microchannel. **d)** First, the microstructures are demolded using the shape memory effect. This step is important to demold the undercuts of the microstructures. **e)** The flat shape memory mold is demolded along the channel direction. **f)** Photographs of the PDMS device with microlenses on the helical microchannel.

temporary programmed microstructures as a spacer allows the casting of microchannels with microstructures on the sidewalls as a solid chip reducing the risk of leakage drastically. The core is thereby completely covered with the casted polymer. Without triggering the shape memory effect, the microstructures are undercuts in the channels sidewalls which makes a demolding of the core impossible. Using the shape memory effect to demold the microstructures before demolding the core enables the fabrication of microstructured microchannels in solid chips. By using the recovery rates investigated in Fig. 6.2, one can also combine temporary programmed microstructures with thermoshaping as proposed before. The result of combining both thermoshaping and temporarily programmed structures is shown in Fig. 6.9. The channels core was temporarily structured with microlenses and thermoshaped in a helical shape. After the PDMS around the channel was cured, the microlenses were demolded by triggering the shape memory effect. Afterwards, the channels core could be demolded and released like any flat core.

6.5. Conclusion and Outlook

Micro- and nanostructured smart molds from shape memory polymers can be used for well established replication processes and offer a new generation of mechanically active molds. To demonstrate this concept, the shape memory effect of two micro- and nanostructured polymers, Tecoflex® EG 72D and Tecoplast® TP 470 was analyzed. The shape memory effect of these polymers allows a precise control of the recovery speed and localized recovery of a permanent shape.

Molds fabricated from shape memory polymers can withstand mechanical stresses and strains during hot embossing and are suitable for casting. For replication of micro- and nanostructures on curved surfaces, the thermoshaping of molds is introduced. Thermoshaping is a two step process. In the first step, a mold is fabricated out of a shape memory polymer with permanently structured or temporarily programmed micro- and nanostructures. In the second step, the flat mold is reprogrammed to a temporary shape at $T = T_{\text{trans}}$ and fixed by holding the temporary shape while cooling down to room temperature. The reprogramming is done by retaining the micro- and nanostructures. In this second step, the initially flat mold can be adapted to curved surfaces, folded around angles or partially elongated. For the later demolding, the mold can be transformed into its permanent shape and demold itself or parts of it. Using dual-shape memory[201] or triple shape memory[74] polymers, the demolding can be stepwise controlled, for instance for sequential demolding of different surfaces of polymer devices.

Additionally, preliminary results using a tip of a scanning thermal microscope to locally heat the shape memory polymer show a high spatial confinement of the restoring to the permanent shape, as shown in Fig. 6.4. The preliminary experiments where carried out in ambient conditions with commercially available heated probes which have some disadvantages compared to the tips used in Chap. 5. They have a poor temperature resolution and also relatively large tip radii of a few hundred of nanometers. They also show a high temperature dependent pre-bending which complicates a precise force control. As the experiments are carried out in ambient conditions, also a high percentage of the heat dissipated at the tip might be transported via air conductance into the shape memory polymer. However, using specifically designed tips for quantitative thermal analysis like used in Chap. 5 can improve the control of the experimental parameters significantly and, therefore, allow a precise addressing of single structural features.

Nevertheless, as the programmed temporary shape was a thinned polymeric foil, pins could be grown locally from the foil. This could also potentially be used on pre-structured foils which had been temporary programmed with a flat surface. With this technique, specific features can be switched on, for instance to locally change the wetting behavior in fluid handling devices.[208]

A patent application has been submitted that refers to the application of shape memory materials for the fabrication of configurable molds and the replication technique therewith. T. Meier, M. Schneider, M. Worgull and H. Hölscher: Patent Application 102013022181.1 (2013).

A shortened version of this chapter will be submitted as the article "Self-Demolding and Healing Micro- and Nanoimprint Molds From Shape Memory Polymers". T. Meier, J. Bur, M. Schneider, M. Worgull, and H. Hölscher: Journal of Micromechanics and Microengineering (2014).

7. Conclusion and Outlook

Modern micro- and nanotechnology offers a huge variety of new applications which influence our everyday life. The key to this technology can be found in the ability to observe matter on these scales. The study of features and material properties on these scales can be utilized for the implementation of new features into new devices. This work has shown approaches to push limits of microscopy techniques towards more versatile instruments. Integrated sensor concepts by magnetoresistive and thermoresistive sensing were applied to locally probe material properties in nanoscale systems and to develop a new fabrication method for application inspired micro- and nanostructures.

As conventional optical microscopy is limited by diffraction, scanning probe microscopy offers higher resolutions. Compared to optical microscopes, however, the field of view is very small. To combine the advantages of the high resolution of a scanning probe microscope with large area scanning, a new atomic force microscope with a nested scanner design was developed. This unique microscope benefits from two independent scanners, one for a large scan area of $800 \times 800\,\mu m^2$ and one for high spatial resolution with a scan area of $5 \times 5\,\mu m^2$. The high resolution scanner, which is placed on top of the large area scanner, enables high resolution imaging of smallest feature sizes due to the remarkable stability of the large area scanner.

Additionally, the instrument was designed to be operated with both, a conventional laser beam deflection setup and self-sensing cantilevers based on magnetoresistive sensing. This concept of applying magnetic tunneling junctions with magnetostrictive electrodes as strain sensitive transducers to atomic force microscopy cantilevers as demonstrated here, showed sufficient sensitivity for usage in atomic force microscopy cantilevers. Furthermore, this first study of the so-called TMR cantilevers, showed a higher sensitivity than specifically optimized piezoresistive and piezoelectric transducers used in atomic force microscopy before. Cantilevers equipped with these sensing elements were used in the most relevant operational modes of atomic force microscopes and performed well in both static contact mode and dynamic modes, such as amplitude and frequency modulation. Hence, magnetoresistive sensing showed remarkable results on imaging atomic step-edges and self-assembled monolayers. Especially for tip scanning and compact, low cost scanning probe microscopes, magnetoresistive sensing is a valuable alternative to optical laser beam deflection setups. Optimizing the measurement electronics as well as of the magnetic tunneling junctions might further improve both strain sensitivity and signal-to-noise ratio. Additionally, different concepts of an integrated magnetic bias field in the magnetic tunneling junctions might be investigated. Finally, this sensor concept might be advanced to other environments like liquids and vacuum, where optical read-out methods are already inferior to self-sensing approaches.

As magnetoresistive sensing can help to simplify instrumentation, cantilevers with integrated sensing elements were also used to improve instrumentation for sensing a specific tip-sample interaction. Using thermoresistive sensors, heat flux between the cantilever's tip and the sample was measured with high precision and enabled thermal conductance measurements on self-assembled monolayers of linear alkane chains as a function of their chain length. For these molecules, there are competing models of heat transport along the chains. The measurements presented here help to shed light on this technologically relevant group of molecules. The quantitative analysis enabled by the thermal force mapping technique allowed classical transport models to be ruled out at these scales. Moreover, signatures of phonon localization and interference effects were observed as the dominant transport mechanism on these scales. For future experiments, this reliable technique can be used for systematic and quantitative investigations on thermal transport along molecular chains. As the molecules tested here can be modified easily and are commercially available in a huge variety, the influence of side-groups on molecular chains, variations of the molecular backbone or coupling strength to thermal reservoirs can be investigated by this new technique.

Using the scanning probe microscopy to characterize the shape memory effect of two mechanically active polymers, allowed the observation of nanoscale recovery from temporary programmed shapes to permanent structures in a time and space resolved manner. Using this insights to the fabrication of micro- and nanostructured molds from shape memory polymers enabled the fabrication of smart molds for micro-and nanoreplication. These molds hold multiple advantages compared to conventional polymeric molds. As they showed excellent healing properties of major deformations of their structural features, this was utilized to extend their lifetime. Furthermore, by introducing thermoshaping, shape memory molds have been used for the replication of micro- and nanostructured curved surfaces on bulk materials. Utilizing the shape memory effect, the mold has been programmed to be self-demolding and allowed the fabrication and replication of polymeric devices which could only be demolded by triggering the self-demolding because of undercuts in the structures. Additionally, preliminary results of localized heating by scanning thermal microscopy can be used for a concept of nanostructures on demand.

A. List of Publications

A.1. Patents

1. T. Meier, M. Schneider, M. Worgull, H. Hölscher. *Formeinsatz aus Formgedächtnismaterial und Replikationsverfahren (Replication Process and Shape Memory Material Mold Insert)*, PATENT APPLICATION **102013022181.1** (2013)

A.2. Articles

1. A. Tavassolizadeh, T. Meier, K. Rott, G. Reiss, E. Quandt, H. Hölscher, D. Meyners. *Self-Sensing Atomic Force Microscopy Cantilevers Based on Tunnel Magnetoresistance Sensors*, Applied Physics Letters, **102**, 153104 (2013)

2. M. Worgull, M. Schneider, M. Röhrig, T. Meier, M. Heilig, A. Kolew, K. Feit, H. Hölscher, J. Leuthold. *Hot Embossing and Thermoforming of Biodegradable Three-Dimensional Wood Structures*, RSC Advances, **3**, 20060-20064 (2013)

3. T. Meier, F. Menges, P. Nirmalray, H. Hölscher, H. Riel, B. Gotsmann. *Length-Dependent Thermal Transport Along Molecular Chains*, Physical Review Letters (2014)

4. M. Worgull, M. Reinhard, M. Röhrig, M. Schneider, T. Meier, M. Heilig, J. Leuthold, H. Hölscher. *Morphing of Nano- and Microstructures: Switching of Functional Surfaces Using Shape Memory Polymers*, to be submitted

5. T. Meier, J. Bur, M. Schneider, M. Worgull, H. Hölscher. *Shape-Memory-Molds for Micro- and Nanoreplication*, to be submitted

6. T. Meier, A. Förste, A. Tavasollizadeh, K. Rott, G. Reiss, E. Quandt, D. Meyners, R. Gröger, T. Schimmel, H. Hölscher. *A Novel Scanning Probe Microscope for Self-Sensing Cantilevers Using a Nested Dual Large Area Scanner*, to be submitted

A.3. Conference Contributions (Oral)

1. T. Meier, A. Tavassolizadeh, E. Quandt, D. Meyners, H. Hölscher. *Magnetoresistive Tunneling Structures With Magnetostrictive Electrodes as Sensors for Atomic Force Microscopy*, 76. JAHRESTAGUNG DER DPG, Berlin, Germany, March 25th - 30th (2012)

2. T. Meier, A. Tavassolizadeh, K. Rott, G. Reiss, E. Quandt, D. Meyners, H. Hölscher. *Self-Sensing Atomic Force Microscopy Cantilevers Based on Tunnel Magneto Resistance*, 10th INTERNATIONAL WORKSHOP ON NANOMECHANICAL SENSING. Stanford, USA, May 1st - 3rd (2013)

3. T. Meier, F. Menges, P. Nirmalraj, H. Hölscher, H. Riel, B. Gotsmann. *Thermal Conductance of Self-Assembled Monolayers Using Scanning Thermal Microscopy*, MRS FALL MEETING & EXHIBIT, Boston, USA, December 1st - 6th (2013)

A.4. Conference Contributions (Poster)

1. T. Meier, Ö. Ünverdi, J.-E. Schmutz, H. Hölscher. *Measuring Wear by Combining Friction Force and Dynamic Force Microscopy*, 2nd INTERNATIONAL WORKSHOP ON ADVANCED ATOMIC FORCE MICROSCOPY TECHNIQUES, Karlsruhe, Germany, February 28th - March 1st (2011)

2. T. Meier, Ö. Ünverdi, J.-E. Schmutz, H. Hölscher. *Combining Friction Force and Dynamic Force Microscopy for Wear Measurements*, 75. JAHRESTAGUNG DER DPG, Dresden, Germany, March 13th - 18th (2011)

3. T. Meier, Ö. Ünverdi, J.-E. Schmutz, H. Hölscher. *Measuring Wear by combining Friction Force and Dynamic Force Microscopy*, 14th INTERNATIONAL CONFERENCE ON NON-CONTACT ATOMIC FORCE MICROSCOPY, Lindau at Lake Constance, Germany, September 18th - 22th (2011)

4. T. Meier, A. Tavassolizadeh, E. Quandt, D. Meyners, H. Hölscher. *Atomic Force Microscopy Cantilevers with Magnetoresistive Sensing*, 3rd INTERNATIONAL WORKSHOP ON ADVANCED ATOMIC FORCE MICROSCOPY TECHNIQUES, Karlsruhe, Germany, March 5th - 6th (2012)

5. T. Meier, A. Tavassolizadeh, K. Rott, G. Reiss, E. Quandt, D. Meyners, H. Hölscher. *Atomic Force Microscopy Cantilevers with Magnetoresistive Sensing*, 4th INTERNATIONAL WORKSHOP ON ADVANCED ATOMIC FORCE MICROSCOPY TECHNIQUES, Karlsruhe, Germany, March 4th - 5th (2013)

6. T. Meier, F. Menges, P. Nirmalraj, H. Hölscher, H. Riel, B. Gotsmann. *Length-Dependent Thermal Transport Along Molecular Chains*, 5th INTERNATIONAL WORKSHOP ON ADVANCED SCANNING PROBE MICROSCOPY TECHNIQUES, Karlsruhe, Germany, February 24th - 25th (2014)

A.5. Scientific Talks at Other Institutions

1. T. Meier, *Scanning Probe Microscopy Using Self-Sensing Cantilevers*, Fritz-Haber-Institute - Department of Chemical Physics, Berlin, Germany, December 13th (2013)

2. T. Meier, H. Hölscher, *Design and Instrumentation for Self-Sensing Atomic Force Microscopy Cantilevers*, Institute of Industrial Science (IIS) - the University of Tokyo, Tokyo, Japan, July 12th (2013)

3. T. Meier, H. Hölscher, *Design and Instrumentation for Self-Sensing Atomic Force Microscopy Cantilevers*, Santa Barbara, USA, Asylum Research Inc., May 6th (2013)

4. T. Meier, A. Tavassolizadeh, D. Meyners, H. Hölscher. *Magnetoresistive Tunneling Structures With Magnetostrictive Electrodes as Sensors for the Atomic Force Microscopy*, IBM Research - Zurich, Rüschlikon, Switzerland, September 12th (2012)

List of Figures

List of Tables

Bibliography

[1] R. P. Feynman. There's Plenty of Room at the Bottom. *Engineering and Science*, 23:22–36, 1960.

[2] E. Abbe. *Die Lehre von der Bildentstehung im Mikroskop*. Vieweg, Braunschweig, 1910.

[3] G. Binnig, H. Rohrer, Ch. Gerber, and E. Weibel. Surface Studies by Scanning Tunneling Microscopy. *Physical Review Letters*, 49:57–61, 1982.

[4] G. Binnig, C. F. Quate, and Ch. Gerber. Atomic Force Microscope. *Physical Review Letters*, 56:930–933, 1986.

[5] D. M. Eigler and E. K. Schweizer. Positioning Single Atoms With a Scanning Tunnelling Microscope. *Nature*, 344:524–526, 1990.

[6] S. Nishida, D. Kobayashi, T. Sakurada, T. Nakazawa, Y. Hoshi, and H. Kawakatsu. Photothermal Excitation and Laser Doppler Velocimetry of Higher Cantilever Vibration Modes for Dynamic Atomic Force Microscopy in Liquid. *Review of Scientific Instruments*, 79:123703, 2008.

[7] R. Kassies, K. O. van der Werf, A. Lenferrnk, C. N. Hunter, J. D. Olsen, V. Subramaniam, and C. Otto. Combined AFM and Confocal Fluorescence Microscope for Applications in Bio-Nanotechnology. *Journal of Microscopy*, 217:109–116, 2005.

[8] A. Tavassolizadeh, T. Meier, K. Rott, G. Reiss, E. Quandt, H. Hölscher, and D. Meyners. Self-Sensing Atomic Force Microscopy

Cantilevers Based on Tunnel Magnetoresistance Sensors. *Applied Physics Letters*, 102:153104, 2013.

[9] B. Gotsmann, M. Lanz, A. Knoll, and U. Dürig. *Nanotechnology*, chapter Nanoscale Thermal and Mechanical Interaction Studies using Heatable Probes, page 043122. Wiley-VCH, 2010.

[10] U. Drechsler, N. Bürer, M. Despont, U. Dürig, B. Gotsmann, F. Robin, and P. Vettiger. Cantilevers With Nano-Heaters for Thermomechanical Storage Application. *Microelectronic Engineering*, 67:397–404, 2003.

[11] A. Majumdar. Scanning Thermal Microscopy. *Annual Review of Materials Science*, 29:505–585, 1999.

[12] F. Menges, H. Riel, A. Stemmer, and B. Gotsmann. Quantitative Thermometry of Nanoscale hot Spots. *Nano Letters*, 12:596–601, 2012.

[13] F. Menges, H. Riel, A. Stemmer, C. Dimitrakopoulos, and B. Gotsmann. Thermal Transport Into Graphene Through Nanoscopic Contacts. *Physical Review Letters*, 111:205901, 2013.

[14] E. Schrödinger. Zur Dynamik elastisch gekoppelter Punktsysteme. *Annalen der Physik*, 349:916–934, 1914.

[15] R. Peierls. Zur kinetischen Theorie der Wärmeleitung in Kristallen. *Annalen der Physik*, 395:1055–1101, 1929.

[16] N. Taniguchi. On the Basic Concept of Nano-Technology. In *Proceedings of International Conference on Production Engineering Tokyo*, 1974.

[17] K. E. Drexler. Molecular Engineering: An Approach to the Development of General Capabilities for Molecular Manipulation. *Pro-

ceedings of the National Academy of Sciences of the United States of America, 78:5275–5278, 1981.

[18] L. Gross, F. Mohn, N. Moll, P. Liljeroth, and G. Meyer. The Chemical Structure of a Molecule Resolved by Atomic Force Microscopy. *Science*, 325:1110–1114, 2009.

[19] F. J. Giessibl. Atomic Resolution of the Silicon (111)-(7x7) Surface by Atomic Force Microscopy. *Science*, 267:68–71, 1995.

[20] H. J. Butt, B. Cappella, and M. Kappl. Force Measurements With the Atomic Force Microscope: Technique, Interpretation and Applications. *Surface Science Reports*, 59:1–152, 2005.

[21] G. Meyer and N. M. Amer. Novel Optical Approach to Atomic Force Microscopy. *Applied Physics Letters*, 53:1045–1047, 1988.

[22] J. Freund, J. Halbritter, and J. K. H. Hörber. How dry are Dried Samples? Water Adsorption Measured by STM. *Microscopy Research and Technique*, 44:327–338, 1999.

[23] T. Stifter, O. Marti, and B. Bhushan. Theoretical Investigation of the Distance Dependence of Capillary and van der Waals Forces in Scanning Force Microscopy. *Physical Review B*, 62:13667–13673, 2000.

[24] J. N. Israelachvili. *Intermolecular and Surface Forces*. Academic Press, 1995.

[25] H. C. Hamaker. The London - van der Waals Attraction Between Spherical Particles. *Physica*, 4:1058–1072, 1937.

[26] W. Pauli. Über den Einfluß der Geschwindigkeitsabhängigkeit der Elektronenmasse auf den Zeemaneffekt. *Zeitschrift für Physik*, 31:373–385, 1925.

[27] W. Pauli. Über den Zusammenhang des Abschlusses der Elektronen-gruppen im Atom mit der Komplexstruktur der Spektren. *Zeitschrift für Physik*, 31:765–783, 1925.

[28] G. E. Uhlenbeck and S. Goudsmit. Spinning Electrons and the Structure of Spectra. *Nature*, 117:264–265, 1926.

[29] G. E. Uhlenbeck and S. Goudsmit. Ersetzung der Hypothese vom unmechanischen Zwang durch eine Forderung bezüglich des inneren Verhaltens jedes einzelnen Elektrons. *Die Naturwissenschaften*, 13:953–954, 1925.

[30] H. Hertz. Über die Berührung fester elastischer Körper. *Journal für die reine und angewandte Mathematik*, 1882:156–171, 1882.

[31] L. D. Landau and E. M. Lifschitz. *Lehrbuch der Theoretischen Physik, Band VII: Elastizitätstheorie*. Akademie-Verlag, 1975.

[32] K. L. Johnson. *Contact Mechanics*. Cambridge University Press, 2003.

[33] K. L. Johnson, K. Kendall, and A. D. Roberts. Surface Energy and the Contact of Elastic Solids. *Proceedings of the Royal Society of London. A. Mathematical and Physical Sciences*, 324:301–313, 1971.

[34] B. V. Derjaguin, V. M. Muller, and Y. P. Toporov. Effect of Contact Deformations on the Adhesion of Particles. *Journal of Colloid and Interface Science*, 53:314–326, 1975.

[35] U. D. Schwarz. A Generalized Analytical Model for the Elastic Deformation of an Adhesive Contact Between a Sphere and a Flat Surface. *Journal of Colloid and Interface Science*, 261:99–106, 2003.

[36] D. Maugis. Adhesion of Spheres: The JKR-DMT Transition Using a Dugdale Model. *Journal of Colloid and Interface Science*, 150:243–269, 1992.

[37] E. Gacoin, C. Fretigny, A. Chateauminois, A. Perriot, and E. Barthel. Measurement of the Mechanical Properties of Thin Films Mechanically Confined Within Contacts. *Tribology Letters*, 21:245–252, 2006.

[38] B. Bhushan and W. Peng. Contact Mechanics of Multilayered Rough Surfaces. *Applied Mechanics Reviews*, 55:435–480, 2002.

[39] S. J. Cole and R. S. Sayles. A Numerical Model for the Contact of Layered Elastic Bodies With Real Rough Surfaces. *Journal of Tribology*, 114:334–340, 1992.

[40] U. D. Schwarz, H. Hölscher, and R. Wiesendanger. Atomic Resolution in Scanning Force Microscopy: Concepts, Requirements, Contrast Mechanisms, and Image Interpretation. *Physical Review B*, 62:13089–13097, 2000.

[41] N. A. Burnham and R. J. Colton. Measuring the Nanomechanical Properties and Surface Forces of Materials Using an Atomic Force Microscope. *Journal of Vacuum Science & Technology A: Vacuum, Surfaces, and Films*, 7:2906–2913, 1989.

[42] C. M. Mate, G. M. McClelland, R. Erlandsson, and S. Chiang. Atomic-Scale Friction of a Tungsten Tip on a Graphite Surface. *Physical Review Letters*, 59:1942–1945, 1987.

[43] Y. Martin, C. C. Williams, and H. K. Wickramasinghe. Atomic Force Microscope - Force Mapping and Profiling on a sub 100-Å Scale. *Journal of Applied Physics*, 61:4723–4729, 1987.

[44] H. Hölscher, D. Ebeling, and U. D. Schwarz. Theory of Q-Controlled Dynamic Force Microscopy in air. *Journal of Applied Physics*, 99:084311, 2006.

[45] U. Dürig. Interaction Sensing in Dynamic Force Microscopy. *New Journal of Physics*, 2:5.1–5.12, 2000.

[46] J. E. Sader, T. Uchihashi, M. J. Higgins, A. Farrell, Y. Nakayama, and S. P. Jarvis. Quantitative Force Measurements Using Frequency Modulation Atomic Force Microscopy - Theoretical Foundations. *Nanotechnology*, 16:S94–S101, 2005.

[47] J. P. Cleveland, B. Anczykowski, A. E. Schmid, and V. B. Elings. Energy Dissipation in Tapping-Mode Atomic Force Microscopy. *Applied Physics Letters*, 72:2613–2615, 1998.

[48] R. Garcia, C. J. Gómez, N. F. Martinez, S. Patil, C. Dietz, and R. Magerle. Identification of Nanoscale Dissipation Processes by Dynamic Atomic Force Microscopy. *Physical Review Letters*, 97:016103, 2006.

[49] T. Sulchek, G. G. Yaralioglu, C. F. Quate, and S. C. Minne. Characterization and Optimization of Scan Speed for Tapping-Mode Atomic Force Microscopy. *Review of Scientific Instruments*, 73:2928–2936, 2002.

[50] G. M. Whitesides and J. C. Love. The art of Building Small. *Scientific American*, September:38–47, 2001.

[51] C. Huang, M. Moosmann, J. Jin, T. Heiler, S. Walheim, and T. Schimmel. Polymer Blend Lithography: A Versatile Method to Fabricate Nanopatterned Self-Assembled Monolayers. *Beilstein Journal of Nanotechnology*, 3:620–628, 2012.

[52] L. Chi. *Nanotechnology*, volume Nanostructured surfaces. Wiley-VCH, 2010.

[53] W. Menz, J. Mohr, and O. Paul. *Mikrosystemtechnik für Ingenieure*. Wiley-VCH, 2005.

[54] V. Saile, editor. *LiGA and its Applications*. Wiley-VCH, 2009.

[55] D. Pires, J. L. Hedrick, A. De Silva, J. Frommer, B. Gotsmann, H. Wolf, M. Despont, U. Duerig, and A. W. Knoll. Nanoscale Three-Dimensional Patterning of Molecular Resists by Scanning Probes. *Science*, 328:732–735, 2010.

[56] A. Waldbaur, H. Rapp, K. Länge, and B. E. Rapp. Let There be Chip - Towards Rapid Prototyping of Microfluidic Devices One-Step Manufacturing Processes. *Analytical Methods*, 3:2681–2716, 2011.

[57] M. Worgull. *Hot Embossing - Theory and Technology of Micro Replication*. William Andrew, Oxford, 2009.

[58] P. J. Flory. Thermodynamics of High Polymer Solutions. *The Journal of Chemical Physics*, 10:51–61, 1942.

[59] S. C. Thickett, A. Harris, and C. Neto. Interplay Between Dewetting and Layer Inversion in Poly(4-vinylpyridine)/Polystyrene Bilayers. *Langmuir*, 26:15989–15999, 2010.

[60] S. Y. Heriot and R. A. L. Jones. An Interfacial Instability in a Transient Wetting Layer Leads to Lateral Phase Separation in Thin Spin-Cast Polymer-Blend Films. *Nature Materials*, 4:782–786, 2005.

[61] G. Reiter. Dewetting of Thin Polymer Films. *Physical Review Letters*, 68:75–78, 1992.

[62] G. Reiter. Unstable Thin Polymer Films: Rupture and Dewetting Processes. *Langmuir*, 9:1344–1351, 1993.

[63] P. Mansky, Y. Liu, E. Huang, T. P. Russell, and C. Hawker. Controlling Polymer-Surface Interactions With Random Copolymer Brushes. *Science*, 275:1458–1460, 1997.

[64] S. Walheim, E. Schäffer, J. Mlynek, and U. Steiner. Nanophase-Separated Polymer Films as High-Performance Antireflection Coatings. *Science*, 283:520–522, 1999.

[65] A. Budkowski, A. Bernasik, P. Cyganik, J. Raczkowska, B. Penc, B. Bergues, K. Kowalski, J. Rysz, and J. Janik. Substrate-Determined Shape of Free Surface Profiles in Spin-Cast Polymer Blend Films. *Macromolecules*, 36:4060–4067, 2003.

[66] W. Madej, A. Budkowski, J. Raczkowska, and J. Rysz. Breath Figures in Polymer and Polymer Blend Films Spin-Coated in dry and Humid Ambience. *Langmuir*, 24:3517–3524, 2008.

[67] M. Heckele, W. Bacher, and K. D. Müller. Hot Embossing - The Molding Technique for Plastic Microstructures. *Microsystem Technologies*, 4:122–124, 1998.

[68] D. M. Cao and W. J. Meng. Microscale Compression Molding of Al With Surface Engineered LiGA Inserts. *Microsystem Technologies*, 10:662–670, 2004.

[69] M. Takahashi, Y. Murakoshi, R. Maeda, and K. Hasegawa. Large Area Micro Hot Embossing of Pyrex Glass with GC Mold Machined by Dicing. *Microsystem Technologies*, 13:379–384, 2007.

[70] G. Kumar, H. X. Tang, and J. Schroers. Nanomoulding With Amorphous Metals. *Nature*, 457:868–872, 2009.

[71] P. C. Vella, S. S. Dimov, A. Kolew, E. Minev, K. Popov, F. Lacan, C. Griffiths, H. Hirshy, and S. Scholz. Bulk Metallic Glass Based Tool - Making Process Chain for Micro- and Nano- Replication. In *Proceedings of the 9th International Conference on Multi-Material Micro Manufacture*, 2012.

[72] I. S. Kolesov and H. J. Radusch. Multiple Shape-Memory Behavior and Thermal-Mechanical Properties of Peroxide Cross-Linked Blends of Linear and Short-Chain Branched Polyethylenes. *Express Polymer Letters*, 2:461–473, 2008.

[73] T. Xie. Tunable Polymer Multi-Shape Memory Effect. *Nature*, 464:267–270, 2010.

[74] M. Behl and A. Lendlein. Triple-Shape Polymers. *Journal of Materials Chemistry*, 20:3335–3345, 2010.

[75] C. Megnin, J. Barth, and M. Kohl. A Bistable SMA Microvalve for 3/2-way Control. *Sensors and Actuators A: Physical*, 188:285–291, 2012.

[76] A. Lendlein and S. Kelch. Shape-Memory Polymers. *Angewandte Chemie International Edition*, 41:2034–2057, 2002.

[77] M. Heuchel, J. Cui, K. Kratz, H. Kosmella, and A. Lendlein. Relaxation Based Modeling of Tunable Shape Recovery Kinetics Observed Under Isothermal Conditions for Amorphous Shape-Memory Polymers. *Polymer*, 51:6212–6218, 2010.

[78] A. Lendlein, M. Behl, B. Hiebl, and C. Wischke. Shape-Memory Polymers as a Technology Platform for Biomedical Applications. *Expert Review of Medical Devices*, 7:357–379, 2010.

[79] J. Cui, K. Kratz, and A. Lendlein. Adjusting Shape-Memory Properties of Amorphous Polyether Urethanes and Radio-Opaque Composites Thereof by Variation of Physical Parameters During Programming. *Smart Materials and Structures*, 19:065019, 2010.

[80] J. Cui, K. Kratz, M. Heuchel, B. Hiebl, and A. Lendlein. Mechanically Active Scaffolds From Radio-Opaque Shape-Memory Polymer-Based Composites. *Polymers for Advanced Technologies*, 22:180–189, 2011.

[81] S. Reddy, E. Arzt, and A. del Campo. Bioinspired Surfaces With Switchable Adhesion. *Advanced Materials*, 19:3833–3837, 2007.

[82] L. G. Carrascosa, M. Moreno, M. Álvarez, and L. M. Lechuga. Nanomechanical Biosensors: A new Sensing Tool. *TrAC Trends in Analytical Chemistry*, 25:196–206, 2006.

[83] R. A. Oliver. Advances in AFM for the Electrical Characterization of Semiconductors. *Reports on Progress in Physics*, 71:076501, 2008.

[84] P. Vettiger, G. Cross, M. Despont, U. Drechsler, U. Dürig, B. Gotsmann, W. Haberle, M. A. Lantz, H. E. Rothuizen, R. Stutz, and G. K. Binnig. The "Millipede" - Nanotechnology Entering Data Storage. *IEEE Transactions on Nanotechnology*, 1:39–55, 2002.

[85] O. Marti, J. Colchero, and J. Mlynek. Combined Scanning Force and Friction Microscopy of Mica. *Nanotechnology*, 1:141–144, 1990.

[86] K. Karrai and R. D. Grober. Piezoelectric Tip-Sample Distance Control for Near Field Optical Microscopes. *Applied Physics Letters*, 66:1842–1844, 1995.

[87] F. J. Giessibl. Atomic Resolution on Si(111)-(7 x 7) by Noncontact Atomic Force Microscopy With a Force Sensor Based on a Quartz Tuning Fork. *Applied Physics Letters*, 76:1470–1472, 2000.

[88] S. Hembacher, F. J. Giessibl, and J. Mannhart. Evaluation of a Force Sensor Based on a Quartz Tuning Fork for Operation at low Temperatures and Ultrahigh Vacuum. *Applied Surface Science*, 188:445–449, 2002.

[89] W. Allers, A. Schwarz, U. D. Schwarz, and R. Wiesendanger. A Scanning Force Microscope with Atomic Resolution in Ultrahigh Vacuum and at low Temperatures. *Review of Scientific Instruments*, 69:221–225, 1998.

[90] H. J. Hug, B. Stiefel, P. J. A. van Schendel, A. Moser, S. Martin, and H. J. Güntherodt. A low Temperature Ultrahigh Vaccum Scanning

Force Microscope. *Review of Scientific Instruments*, 70:3625–3640, 1999.

[91] B. Gotsmann and M. A. Lantz. Quantized Thermal Transport Across Contacts of Rough Surfaces. *Nature Materials*, 12:59–65, 2012.

[92] M. Tortonese, H. Yamada, R. C. Barrett, and C. F. Quate. Atomic Force Microscopy Using a Piezoresistive Cantilever. In *1991 International Conference on Solid-State Sensors and Actuators, 1991. Digest of Technical Papers, TRANSDUCERS '91.*, 1991.

[93] J. Tansock and C. C. Williams. Force Measurement With a Piezoelectric Cantilever in a Scanning Force Microscope. *Ultramicroscopy*, 42-44:1464–1469, 1992.

[94] M. Tortonese, R. C. Barrett, and C. F. Quate. Atomic Resolution With an Atomic Force Microscope Using Piezoresistive Detection. *Applied Physics Letters*, 62:834–836, 1993.

[95] R. Linnemann, T. Gotszalk, L. Hadjiiski, and I. W. Rangelow. Characterization of a Cantilever With an Integrated Deflection Sensor. *Thin Solid Films*, 264:159–164, 1995.

[96] R. Linnemann, T. Gotszalk, I. W. Rangelow, P. Dumania, and E. Oesterschulze. Atomic Force Microscopy and Lateral Force Microscopy Using Piezoresistive Cantilevers. *Journal of Vacuum Science & Technology B: Microelectronics and Nanometer Structures*, 14:856–860, 1996.

[97] T. Gotszalk, P. Grabiec, and I. W. Rangelow. Piezoresistive Sensors for Scanning Probe Microscopy. *Ultramicroscopy*, 82:39–48, 2000.

[98] J. C. Doll and B. L. Pruitt. Design of Piezoresistive Versus Piezoelectric Contact Mode Scanning Probes. *Journal of Micromechanics and Microengineering*, 20:095023, 2010.

[99] G. Neubauer, S. R. Cohen, G. M. McClelland, D. Horne, and C. M. Mate. Force Microscopy With a Bidirectional Capacitance Sensor. *Review of Scientific Instruments*, 61:2296–2308, 1990.

[100] S. A. Miller, K. L. Turner, and N. C. MacDonald. Microelectrome-chanical Scanning Probe Instruments for Array Architectures. *Review of Scientific Instruments*, 68:4155–4162, 1997.

[101] A. J. Waddie, M. R. Taghizadeh, J. Mohr, V. Piotter, Ch. Mehne, A. Stuck, E. Stijns, and H. Thienpont. Design, Fabrication and Replication of Micro-Optical Components for Educational Purposes within the Network of Excellence in Micro-Optics (NEMO). In *Proceedings of the SPIE (Photonics Europe 2006)*, 2006.

[102] J. G. Zhu. New Heights for Hard Disk Drives. *Materials Today*, 6:22–31, 2003.

[103] T. S. Plaskett, P. P. Freitas, N. P. Barradas, M. F. da Silva, and J. C. Soares. Magnetoresistance and Magnetic Properties of NiFe/Oxide/Co Junctions Prepared by Magnetron Sputtering. *Journal of Applied Physics*, 76:6104–6106, 1994.

[104] J. S. Moodera, L. R. Kinder, T. M. Wong, and R. Meservey. Large Magnetoresistance at Room Temperature in Ferromagnetic Thin Film Tunnel Junctions. *Physical Review Letters*, 74:3273–3276, 1995.

[105] J. S. Moodera, E. F. Gallagher, K. Robinson, and J. Nowak. Optimum Tunnel Barrier in Ferromagnetic-Insulator-Ferromagnetic Tunneling Structures. *Applied Physics Letters*, 70:3050–3050, 1997.

[106] R. C. Sousa, J. J. Sun, V. Soares, P. P. Freitas, A. Kling, M. F. da Silva, and J. C. Soares. Large Tunneling Magnetoresistance Enhancement by Thermal Anneal. *Applied Physics Letters*, 73:3288–3290, 1998.

[107] J. S. Moodera, J. Nassar, and G. Mathon. Spin-Tunneling in Ferromagnetic Junctions. *Annual Review of Materials Science*, 29:381–432, 1999.

[108] D. Wang, C. Nordman, J. M. Daughton, Z. Qian, and J. Fink. 70,% TMR at Room Temperature for SDT Sandwich Junctions With CoFeB as Free and Reference Layers. *IEEE Transactions on Magnetics*, 40:2269–2271, 2004.

[109] S. Yuasa, T. Nagahama, A. Fukushima, Y. Suzuki, and K. Ando. Giant Room-Temperature Magnetoresistance in Single-Crystal Fe/MgO/Fe Magnetic Tunnel Junctions. *Nature Materials*, 3:868–871, 2004.

[110] S. S. P. Parkin, C. Kaiser, A. Panchula, P. M. Rice, B. Hughes, M. Samant, and S. H. Yang. Giant Tunnelling Magnetoresistance at Room Temperature with MgO(100) Tunnel Barriers. *Nature Materials*, 3:862–867, 2004.

[111] D. Meyners, T. von Hofe, M. Vieth, M. Ruhrig, S. Schmitt, and E. Quandt. Pressure Sensor Based on Magnetic Tunnel Junctions. *Journal of Applied Physics*, 105:07C914, 2009.

[112] C. Albon, A. Weddemann, A. Auge, K. Rott, and A. Hütten. Tunneling Magnetoresistance Sensors for High Resolutive Particle Detection. *Applied Physics Letters*, 95:023101, 2009.

[113] W. Thomson. On the Electro-Dynamic Qualities of Metals: Effects of Magnetization on the Electric Conductivity of Nickel and of Iron. *Proceedings of the Royal Society of London*, 8:546–550, 1856.

[114] P. Grünberg, R. Schreiber, Y. Pang, M. B. Brodsky, and H. Sowers. Layered Magnetic Structures: Evidence for Antiferromagnetic Coupling of Fe Layers Across Cr Interlayers. *Physical Review Letters*, 57:2442–2445, 1986.

[115] M. Julliere. Tunneling Between Ferromagnetic Films. *Physics Letters A*, 54:225–226, 1975.

[116] T. Miyazaki, T. Yaoi, and S. Ishio. Large Magnetoresistance Effect in 82Ni-Fe/Al-Al$_2$O$_3$/Co Magnetic Tunneling Junction. *Journal of Magnetism and Magnetic Materials*, 98:L7–L9, 1991.

[117] W. H. Butler, X. G. Zhang, T. C. Schulthess, and J. M. MacLaren. Spin-Dependent Tunneling Conductance of Fe-MgO-Fe Sandwiches. *Physical Review B*, 63:054416, 2001.

[118] J. Mathon and A. Umerski. Theory of Tunneling Magnetoresistance of an Epitaxial Fe/MgO/Fe(001) Junction. *Physical Review B*, 63:220403, 2001.

[119] J. G. Simmons. Generalized Formula for the Electric Tunnel Effect Between Similar Electrodes Separated by a Thin Insulating Film. *Journal of Applied Physics*, 34:1793–1803, 1963.

[120] R. J. Soulen, J. M. Byers, M. S. Osofsky, B. Nadgorny, T. Ambrose, S. F. Cheng, P. R. Broussard, C. T. Tanaka, J. Nowak, J. S. Moodera, A. Barry, and J. M. D. Coey. Measuring the Spin Polarization of a Metal With a Superconducting Point Contact. *Science*, 282:85–88, 1998.

[121] D. J. Monsma and S. S. P. Parkin. Spin Polarization of Tunneling Current From Ferromagnet/Al$_2$O$_3$ Interfaces Using Copper-Doped Aluminum Superconducting Films. *Applied Physics Letters*, 77:720–722, 2000.

[122] J. G. Zhu and C. Park. Magnetic Tunnel Junctions. *Materials Today*, 9:36–45, 2006.

[123] R. Landauer. Spatial Variation of Currents and Fields due to Localized Scatterers in Metallic Conduction. *IBM Journal of Research and Development*, 1:223–231, 1957.

[124] S. Yuasa. Giant Tunneling Magnetoresistance in MgO-Based Magnetic Tunnel Junctions. *Journal of the Physical Society of Japan*, 77:031001, 2008.

[125] M. Löhndorf, T. Duenas, M. Tewes, E. Quandt, M. Rührig, and J. Wecker. Highly Sensitive Strain Sensors Based on Magnetic Tunneling Junctions. *Applied Physics Letters*, 81:313–315, 2002.

[126] T. Duenas, A. Sehrbrock, M. Löhndorf, A. Ludwig, J. Wecker, P. Grünberg, and E. Quandt. Micro-Sensor Coupling Magnetostriction and Magnetoresistive Phenomena. *Journal of Magnetism and Magnetic Materials*, 242-245:1132–1135, 2002.

[127] A. Hubert and R. Schäfer. *Magnetic Domains: The Analysis of Magnetic Microstructures*. Springer, 1998.

[128] H. Kronmüller. *Handbook of Magnetism and Advanced Magnetic Materials*. Wiley-VCH, 2007.

[129] A. Aharoni. *Introduction to the Theory of Ferromagnetism*. Oxford University Press, 2000.

[130] R. C. O'Handley. *Modern Magnetic Materials: Principles and Applications*. Wiley-VCH, 2000.

[131] B. Zhu, C. C. H. Lo, S. J. Lee, and D. C. Jiles. Micromagnetic Modeling of the Effects of Stress on Magnetic Properties. *Journal of Applied Physics*, 89:7009–7011, 2001.

[132] J. Zhang and R. M. White. Voltage Dependence of Magnetoresistance in Spin Dependent Tunneling Junctions. *Journal of Applied Physics*, 83:6512–6514, 1998.

[133] P. O. Chapuis, J. J. Greffet, K. Joulain, and S. Volz. Heat Transfer Between a Nano-tip and a Surface. *Nanotechnology*, 17:2978–2981, 2006.

[134] M. Hinz, O. Marti, B. Gotsmann, M. A. Lantz, and U. Dürig. High Resolution Vacuum Scanning Thermal Microscopy of HfO_2 and SiO_2. *Applied Physics Letters*, 92:043122, 2008.

[135] S. M. Sze and K. K. Ng. *Physics of Semiconductor Devices*. Wiley-VCH, 2007.

[136] H. J. Mamin, B. A. Gurney, D. R. Wilhoit, and V. S. Speriosu. High Sensitivity Spin-Valve Strain Sensor. *Applied Physics Letters*, 72:3220–3222, 1998.

[137] B. A. Gurney, H. J. Mamin, D. Rugar, and V. S. Speriosu. Atomic Force Microscope System With Cantilevers Having Unbiased Spin Valve Magnetoresistive Strain Gauge, 1999. US Patent 5,856,617.

[138] D. R. Sahoo, A. Sebastian, W. Häberle, H. Pozidis, and E. Eleftheriou. Scanning Probe Microscopy Based on Magnetoresistive Sensing. *Nanotechnology*, 22:145501, 2011.

[139] H. Bhaskaran, M. Li, D. Garcia-Sanchez, P. Zhao, I. Takeuchi, and H. X. Tang. Active Microcantilevers Based on Piezoresistive Ferromagnetic Thin Films. *Applied Physics Letters*, 98:013502, 2011.

[140] J. Hayakawa, S. Ikeda, F. Matsukura, H. Takahashi, and H. Ohno. Dependence of Giant Tunnel Magnetoresistance of Sputtered CoFeB/MgO/CoFeB Magnetic Tunnel Junctions on MgO Barrier Thickness and Annealing Temperature. *Japanese Journal of Applied Physics*, 44:L587–L589, 2005.

[141] Y. M. Lee, J. Hayakawa, S. Ikeda, F. Matsukura, and H. Ohno. Giant Tunnel Magnetoresistance and High Annealing Stability in CoFeB / MgO / CoFeB Magnetic Tunnel Junctions With Synthetic Pinned Layer. *Applied Physics Letters*, 89:042506, 2006.

[142] D. Meyners, J. Puchalla, S. Dokupil, M. Löhndorf, and E. Quandt. Magnetoelectronical Sensors for Mechanical Measurements. *ECS Transactions*, 3:223–233, 2007.

[143] S. Ikeda, J. Hayakawa, Y. M. Lee, F. Matsukura, Y. Ohno, T. Hanyu, and H. Ohno. Magnetic Tunnel Junctions for Spintronic Memories and Beyond. *IEEE Transactions on Electron Devices*, 54:991–1002, 2007.

[144] H. Jaffrès, D. Lacour, F. Nguyen Van Dau, J. Briatico, F. Petroff, and A. Vaurès. Angular Dependence of the Tunnel Magnetoresistance in Transition-Metal-Based Junctions. *Physical Review B*, 64:064427, 2001.

[145] A. Gaitas, T. Li, and W. Zhu. A Probe With Ultrathin Film Deflection Sensor for Scanning Probe Microscopy and Material Characterization. *Sensors and Actuators A: Physical*, 168:229–232, 2011.

[146] J. Thaysen, A. Boisen, O. Hansen, and S. Bouwstra. Atomic Force Microscopy Probe With Piezoresistive Read-out and a Highly Symmetrical Wheatstone Bridge Arrangement. *Sensors and Actuators A: Physical*, 83:47–53, 2000.

[147] X. Yu, J. Thaysen, O. Hansen, and A. Boisen. Optimization of Sensitivity and Noise in Piezoresistive Cantilevers. *Journal of Applied Physics*, 92:6296–6301, 2002.

[148] J. Lee and W. P. King. Improved All-Silicon Microcantilever Heaters With Integrated Piezoresistive Sensing. *Journal of Microelectromechanical Systems*, 17:432–445, 2008.

[149] M. Qazi, N. DeRoller, A. Talukdar, and G. Koley. III-V Nitride Based Piezoresistive Microcantilever for Sensing Applications. *Applied Physics Letters*, 99:193508, 2011.

[150] G. Binnig, Ch. Gerber, E. Stoll, T. R. Albrecht, and C. F. Quate. Atomic Resolution With Atomic Force Microscope. *Surface Science*, 189-190:1–6, 1987.

[151] J. H. Kindt, G. E. Fantner, J. B. Thompson, and P. K. Hansma. Automated Wafer-Scale Fabrication of Electron Beam Deposited Tips for Atomic Force Microscopes Using Pattern Recognition. *Nanotechnology*, 15:1131–1134, 2004.

[152] D. Ebeling, B. Eslami, and S. D. J. Solares. Visualizing the Subsurface of Soft Matter: Simultaneous Topographical Imaging, Depth Modulation, and Compositional Mapping With Triple Frequency Atomic Force Microscopy. *ACS Nano*, 7:10387–10396, 2013.

[153] A. Noy, C. H. Sanders, D. V. Vezenov, S. S. Wong, and C. M. Lieber. Chemically-Sensitive Imaging in Tapping Mode by Chemical Force Microscopy: Relationship Between Phase lag and Adhesion. *Langmuir*, 14:1508–1511, 1998.

[154] S. Huxtable, D. G. Cahill, V. Fauconnier, J. O. White, and J. C. Zhao. Thermal Conductivity Imaging at Micrometer-Scale Resolution for Combinatorial Studies of Materials. *Nature Materials*, 3:298–301, 2004.

[155] D. G. Cahill, W. K. Ford, K. E. Goodson, G. D. Mahan, A. Majumdar, H. J. Maris, R. Merlin, and S. R. Phillpot. Nanoscale Thermal Transport. *Journal of Applied Physics*, 93:793–818, 2003.

[156] E. Fermi, J. Pasta, and S. Ulam. Studies of Nonlinear Problems. *Los Alamos Document LA-1940*, 1955.

[157] S. Lepri, R. Livi, and A. Politi. Thermal Conduction in Classical Low-Dimensional Lattices. *Physics Reports*, 377:1–80, 2003.

[158] A. Dhar. Heat Transport in Low-Dimensional Systems. *Advances in Physics*, 57:457–537, 2008.

[159] D. Witt, R. Klajn, P. Barski, and B. A. Grzybowski. *Current Organic Chemistry*. Bentham Science, 2004.

[160] A. Zeira, J. Berson, I. Feldman, R. Maoz, and J. Sagiv. A Bipolar Electrochemical Approach to Constructive Lithography: Metal/Monolayer Patterns via Consecutive Site-Defined Oxidation and Reduction. *Langmuir*, 27:8562–8575, 2011.

[161] A. Ulman. Formation and Structure of Self-Assembled Monolayers. *Chemical Reviews*, 96:1533–1554, 1996.

[162] D. Segal, A. Nitzan, and P. Hänggi. Thermal Conductance Through Molecular Wires. *The Journal of Chemical Physics*, 119:6840–6855, 2003.

[163] M. A. Panzer and K. E. Goodson. Thermal Resistance Between Low-Dimensional Nanostructures and Semi-Infinite Media. *Journal of Applied Physics*, 103:094301, 2008.

[164] L. Hu, L. Zhang, M. Hu, J. S. Wang, B. Li, and P. Keblinski. Phonon Interference at Self-Assembled Monolayer Interfaces: Molecular Dynamics Simulations. *Physical Review B*, 81:235427, 2010.

[165] Z. Ge, D. G. Cahill, and P. V. Braun. Thermal Conductance of Hydrophilic and Hydrophobic Interfaces. *Physical Review Letters*, 96:186101, 2006.

[166] Z. Wang, J. A. Carter, A. Lagutchev, Y. K. Koh, N. H. Seong, D. G. Cahill, and D. D. Dlott. Ultrafast Flash Thermal Conductance of Molecular Chains. *Science*, 317:787–790, 2007.

[167] R. Y. Wang, R. A. Segalman, and A. Majumdar. Room Temperature Thermal Conductance of Alkanedithiol Self-Assembled Monolayers. *Applied Physics Letters*, 89:173113, 2006.

[168] M. D. Losego, M. E. Grady, N. R. Sottos, D. G. Cahill, and P. V. Braun. Effects of Chemical Bonding on Heat Transport Across Interfaces. *Nature Materials*, 11:502–506, 2012.

[169] P. J. O'Brien, S. Shenogin, J. Liu, P. K. Chow, D. Laurencin, P. H. Mutin, M. Yamaguchi, P. Keblinski, and G. Ramanath. Bonding-Induced Thermal Conductance Enhancement at Inorganic Heterointerfaces Using Nanomolecular Monolayers. *Nature Materials*, 12:118–122, 2013.

[170] J. C. Duda, C. B. Saltonstall, P. M. Norris, and P. E. Hopkins. Assessment and Prediction of Thermal Transport at Solid - Self-Assembled Monolayer Junctions. *The Journal of Chemical Physics*, 134:094704, 2011.

[171] D. Schwarzer, P. Kutne, C. Schröder, and J. Troe. Intramolecular Vibrational Energy Redistribution in Bridged Azulene-Anthracene Compounds: Ballistic Energy Transport Through Molecular Chains. *The Journal of Chemical Physics*, 121:1754–1764, 2004.

[172] J. C. Love, L. A. Estroff, J. K. Kriebel, R. G. Nuzzo, and G. M. Whitesides. Self-Assembled Monolayers of Thiolates on Metals as a Form of Nanotechnology. *Chemical Reviews*, 105:1103–1170, 2005.

[173] M. Monchiero, R. Canal, and A. Gonzalez. Power/Performance/Thermal Design-Space Exploration for Multicore Architectures. *IEEE Transactions on Parallel and Distributed Systems*, 19:666–681, 2008.

[174] E. Lortscher, D. Widmer, and B. Gotsmann. Next-Generation Nanotechnology Laboratories With Simultaneous Reduction of all Relevant Disturbances. *Nanoscale*, 5:10542–10549, 2013.

[175] L. Ramin and A. Jabbarzadeh. Effect of Compression on Self-Assembled Monolayers: A Molecular Dynamics Study. *Modelling*

and Simulation in Materials Science and Engineering, 20:085010, 2012.

[176] K. Kurabayashi. Anisotropic Thermal Properties of Solid Polymers. *International Journal of Thermophysics*, 22:277–288, 2001.

[177] S. Shen, A. Henry, J. Tong, R. Zheng, and G. Chen. Polyethylene Nanofibres With Very High Thermal Conductivities. *Nature Nanotechnology*, 5:251–255, 2010.

[178] K. Kurabayashi, M. Asheghi, M. Touzelbaev, and K. E. Goodson. Measurement of the Thermal Conductivity Anisotropy in Polyimide Films. *Journal of Microelectromechanical Systems*, 8:180–191, 1999.

[179] G. Chen. *Nanoscale Energy Transport and Conversion: A Parallel Treatment of Electrons, Molecules, Phonons, and Photons*. MIT-Pappalardo series in mechanical engineering. Oxford University Press, 2005.

[180] D. M. Rowe. *Thermoelectrics Handbook: Macro to Nano*. Taylor & Francis, 2006.

[181] D. Li, Y. Wu, P. Kim, L. Shi, P. Yang, and A. Majumdar. Thermal Conductivity of Individual Silicon Nanowires. *Applied Physics Letters*, 83:2934–2936, 2003.

[182] D. Wiesmann and A. Sebastian. Dynamics of Silicon Micro-Heaters: Modelling and Experimental Identification. In *19th IEEE International Conference on Micro Electro Mechanical Systems, 2006. MEMS*, pages 182–185, 2006.

[183] K. J. Kim and W. P. King. Thermal Conduction Between a Heated Microcantilever and a Surrounding air Environment. *Applied Thermal Engineering*, 29:1631–1641, 2009.

[184] J. E. Schmutz, H. Fuchs, and H. Hölscher. Measuring Wear by Combining Friction Force and Dynamic Force Microscopy. *Wear*, 268:526–532, 2010.

[185] B. Gotsmann and M. A. Lantz. Atomistic Wear in a Single Asperity Sliding Contact. *Physical Review Letters*, 101:125501, 2008.

[186] T. E. Balmer, H. Schmid, R. Stutz, E. Delamarche, B. Michel, N. D. Spencer, and H. Wolf. Diffusion of Alkanethiols in PDMS and its Implications on Microcontact Printing (μCP). *Langmuir*, 21:622–632, 2005.

[187] G. Y. Liu, S. Xu, and Y. Qian. Nanofabrication of Self-Assembled Monolayers Using Scanning Probe Lithography. *Accounts of Chemical Research*, 33:457–466, 2000.

[188] R. W. Carpick and M. Salmeron. Scratching the Surface: Fundamental Investigations of Tribology With Atomic Force Microscopy. *Chemical Reviews*, 97:1163–1194, 1997.

[189] L. Costelle, P. Jalkanen, M. T. Räisänen, L. Lind, R. Nowak, and J. Räisänen. Mechanical Response of Nanometer Thick Self-Assembled Monolayers on Metallic Substrates Using Classical Nanoindentation. *Journal of Applied Physics*, 110:114301, 2011.

[190] E. Delamarche, B. Michel, Ch. Gerber, D. Anselmetti, H. J. Güntherodt, H. Wolf, and H. Ringsdorf. Real-Space Observation of Nanoscale Molecular Domains in Self-Assembled Monolayers. *Langmuir*, 10:2869–2871, 1994.

[191] D. Roy. Crossover From Ballistic to Diffusive Thermal Transport in Quantum Langevin Dynamics Study of a Harmonic Chain Connected to Self-Consistent Reservoirs. *Physical Review E*, 77:062102, 2008.

[192] K. R. Patton and M. R. Geller. Thermal Transport Through a Mesoscopic Weak Link. *Physical Review B*, 64:155320, 2001.

[193] M. Röhrig, M. Thiel, M. Worgull, and H. Hölscher. 3D Direct Laser Writing of Nano- and Microstructured Hierarchical Gecko-Mimicking Surfaces. *Small*, 8:3009–3015, 2012.

[194] M. Röhrig, M. Schneider, G. Etienne, F. Oulhadj, F. Pfannes, A. Kolew, M. Worgull, and H. Hölscher. Hot Pulling and Embossing of Hierarchical Nano- and Micro-Structures. *Journal of Micromechanics and Microengineering*, 23:105014, 2013.

[195] R. H. Siddique, S. Diewald, J. Leuthold, and H. Hölscher. Theoretical and Experimental Analysis of the Structural Pattern Responsible for the Iridescence of Morpho Butterflies. *Optics Express*, 21:14351–14361, 2013.

[196] T. Katoh, R. Tokuno, Y. Zhang, M. Abe, K. Akita, and M. Akamatsu. Micro Injection Molding for Mass Production Using LiGA Mold Inserts. *Microsystem Technologies*, 14:1507–1514, 2008.

[197] A. Kolew, D. Münch, K. Sikora, and M. Worgull. Hot Embossing of Micro and Sub-Micro Structured Inserts for Polymer Replication. *Microsystem Technologies*, 17:609–618, 2011.

[198] N. E. Steidle, M. Schneider, R. Ahrens, M. Worgull, and A. E. Guber. Fabrication of Polymeric Microfluidic Devices With Tunable Wetting Behavior for Biomedical Applications. In *Engineering in Medicine and Biology Society (EMBC), 2013 35th Annual International Conference of the IEEE*, pages 6659–6662, July 2013.

[199] M. Heilig, S. Giselbrecht, A. Guber, and M. Worgull. Microthermoforming of Nanostructured Polymer Films: A new Bonding Method for the Integration of Nanostructures in 3-Dimensional Cavities. *Microsystem Technologies*, 16:1221–1231, 2010.

[200] M. Behl and A. Lendlein. Shape-Memory Polymers. *Materials Today*, 10:20–28, 2007.

[201] T. Xie. Recent Advances in Polymer Shape Memory. *Polymer*, 52:4985–5000, 2011.

[202] Q. Zhang, M. Behl, and A. Lendlein. Shape-Memory Polymers With Multiple Transitions: Complex Actively Moving Polymers. *Soft Matter*, 9:1744–1755, 2013.

[203] K. Ikuta, M. Tsukamoto, and S. Hirose. Shape Memory Alloy Servo Actuator System With Electric Resistance Feedback and Application for Active Endoscope. In *1988 IEEE International Conference on Robotics and Automation*, 1988.

[204] P. Krulevitch, A. P. Lee, P. B. Ramsey, J. C. Trevino, J. Hamilton, and M. A. Northrup. Thin Film Shape Memory Alloy Microactuators. *Journal of Microelectromechanical Systems*, 5:270–282, 1996.

[205] A. Lendlein and R. Langer. Biodegradable, Elastic Shape-Memory Polymers for Potential Biomedical Applications. *Science*, 296:1673–1676, 2002.

[206] Senta Schauer. Herstellung, Programmierung und Charakterisierung von mikrooptischen Gittern aus einem Formgedächnispolymer. Master's thesis, Karlsruhe Institute of Technology, 2013.

[207] L. Shi, J. Zhou, P. Kim, A. Bachtold, A. Majumdar, and P. L. McEuen. Thermal Probing of Energy Dissipation in Current-Carrying Carbon Nanotubes. *Journal of Applied Physics*, 105:104306, 2009.

[208] M. Worgull, M. Schneider, M. Röhrig, T. Meier, M. Heilig, A. Kolew, K. Feit, H. Hölscher, and J. Leuthold. Hot Embossing and Thermoforming of Biodegradable Three-Dimensional Wood Structures. *RSC Advances*, 3:20060–20064, 2013.

Schriften des Instituts für Mikrostrukturtechnik
am Karlsruher Institut für Technologie (KIT)

ISSN 1869-5183

Herausgeber: Institut für Mikrostrukturtechnik

Die Bände sind unter www.ksp.kit.edu als PDF frei verfügbar
oder als Druckausgabe zu bestellen.

Band 1   Georg Obermaier
          Research-to-Business Beziehungen: Technologietransfer durch
          Kommunikation von Werten (Barrieren, Erfolgsfaktoren und
          Strategien). 2009
          ISBN 978-3-86644-448-5

Band 2   Thomas Grund
          Entwicklung von Kunststoff-Mikroventilen im Batch-Verfahren. 2010
          ISBN 978-3-86644-496-6

Band 3   Sven Schüle
          Modular adaptive mikrooptische Systeme in Kombination
          mit Mikroaktoren. 2010
          ISBN 978-3-86644-529-1

Band 4   Markus Simon
          Röntgenlinsen mit großer Apertur. 2010
          ISBN 978-3-86644-530-7

Band 5   K. Phillip Schierjott
          Miniaturisierte Kapillarelektrophorese zur kontinuierlichen Über-
          wachung von Kationen und Anionen in Prozessströmen. 2010
          ISBN 978-3-86644-523-9

Band 6   Stephanie Kißling
          Chemische und elektrochemische Methoden zur Oberflächenbe-
          arbeitung von galvanogeformten Nickel-Mikrostrukturen. 2010
          ISBN 978-3-86644-548-2

Schriften des Instituts für Mikrostrukturtechnik
am Karlsruher Institut für Technologie (KIT)

ISSN 1869-5183

Band 7   Friederike J. Gruhl
          Oberflächenmodifikation von Surface Acoustic Wave (SAW)
          Biosensoren für biomedizinische Anwendungen. 2010
          ISBN 978-3-86644-543-7

Band 8   Laura Zimmermann
          Dreidimensional nanostrukturierte und superhydrophobe
          mikrofluidische Systeme zur Tröpfchengenerierung und
          -handhabung. 2011
          ISBN 978-3-86644-634-2

Band 9   Martina Reinhardt
          Funktionalisierte, polymere Mikrostrukturen für die
          dreidimensionale Zellkultur. 2011
          ISBN 978-3-86644-616-8

Band 10   Mauno Schelb
          Integrierte Sensoren mit photonischen Kristallen auf
          Polymerbasis. 2012
          ISBN 978-3-86644-813-1

Band 11   Daniel Auernhammer
          Integrierte Lagesensorik für ein adaptives mikrooptisches
          Ablenksystem. 2012
          ISBN 978-3-86644-829-2

Band 12   Nils Z. Danckwardt
          Pumpfreier Magnetpartikeltransport in einem Mikroreaktions-
          system: Konzeption, Simulation und Machbarkeitsnachweis. 2012
          ISBN 978-3-86644-846-9

Band 13   Alexander Kolew
          Heißprägen von Verbundfolien für mikrofluidische
          Anwendungen. 2012
          ISBN 978-3-86644-888-9

Schriften des Instituts für Mikrostrukturtechnik
am Karlsruher Institut für Technologie (KIT)

ISSN 1869-5183

Band 14   Marko Brammer
           Modulare Optoelektronische Mikrofluidische Backplane. 2012
           ISBN 978-3-86644-920-6

Band 15   Christiane Neumann
           Entwicklung einer Plattform zur individuellen Ansteuerung von
           Mikroventilen und Aktoren auf der Grundlage eines Phasenüber-
           ganges zum Einsatz in der Mikrofluidik. 2013
           ISBN 978-3-86644-975-6

Band 16   Julian Hartbaum
           Magnetisches Nanoaktorsystem. 2013
           ISBN 978-3-86644-981-7

Band 17   Johannes Kenntner
           Herstellung von Gitterstrukturen mit Aspektverhältnis 100 für die
           Phasenkontrastbildgebung in einem Talbot-Interferometer. 2013
           ISBN 978-3-7315-0016-2

Band 18   Kristina Kreppenhofer
           Modular Biomicrofluidics - Mikrofluidikchips im Baukastensystem
           für Anwendungen aus der Zellbiologie. 2013
           ISBN 978-3-7315-0036-0

Band 19   Ansgar Waldbaur
           Entwicklung eines maskenlosen Fotolithographiesystems zum
           Einsatz im Rapid Prototyping in der Mikrofluidik und zur gezielten
           Oberflächenfunktionalisierung. 2013
           ISBN 978-3-7315-0119-0

Band 20   Christof Megnin
           Formgedächtnis-Mikroventile für eine fluidische Plattform. 2013
           ISBN 978-3-7315-0121-3

Schriften des Instituts für Mikrostrukturtechnik
am Karlsruher Institut für Technologie (KIT)

ISSN 1869-5183